METROPOLIS

METROPOLIS

MAPPING THE CITY

Jeremy Black

Contributing editor Christopher Westhorp

B L O O M S B U R Y

LONDON · NEW DELHI · NEW YORK · SYDNEY

Published by Conway
An imprint of Bloomsbury Publishing Plc
www.bloomsbury.com

50 Bedford Square   1385 Broadway
London   New York
WC1B 3DP   NY 10018
UK   USA

First published 2015

British Library Cataloguing-in-Publication Data
A catalogue record for this book is available from the British Library.

ISBN: HB: 978-1-8448-6220-7

10 9 8 7 6 5 4 3 2 1

Design by Nicola Liddiard at Nimbus Design

Printed in Singapore

To find out more about our authors and books visit *www.bloomsbury.com*.

Here you will find extracts, author interviews, details of forthcoming
events and the option to sign up for our newsletters.

*It is a great pleasure to dedicate this book to Jeannie Forbes
with much love from Sarah and me.*

Author's acknowledgments

As explained in the introduction, cities provide major subjects and topics for mapping, and challenges for effective depiction, and so also with the role of cities in wider historical, geographical and cultural spheres. Cities constitute a major strand in the history of the world and this has been great fun to research and write. I would like to thank Alison Moss and John Lee for the opportunity to do so; and for help with the illustrations, I am most grateful to Jennifer Veall. While thinking about and working on this book, I benefited from the chances to visit Birmingham, Bratislava, Brussels, Bucharest, Budapest, Charleston, Chennai, Chicago, Cochin, Cologne, Colombo, Copenhagen, Havana, Kyoto, London, Mainz, Montreal, New York, Paris, Philadelphia, Quebec, Singapore, Tokyo, Toronto, Vienna and Washington. The opportunity, in giving the 2011 Marc Fitch Lecture on London's history, to test out ideas proved most welcome.

WASHINGTON DC, 1884 Detail from Adolph Sachse's 1883–1884 panorama of the Capitol which can be seen in full on pages 152–153. The White House can be seen in the bottom left hand corner. **p2**

IRISH CITIES A set of maps of the Irish cities of Galway, Dublin, Limerick and Cork, dating from the late 16th century and taken from the *Civitates Orbis Terrarum* by Georg Braun and Franz Hogenberg. **p6**

NAGASAKI A colour woodcut map of the Nagasaki harbour area in 1801 by Yamato-Ya. **p8**

Contents

GALWAYE

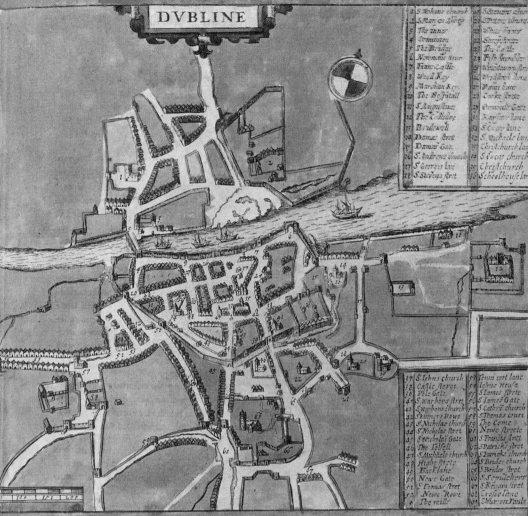

DVBLINE

1	S Johans church	33	S Eunans church
2	S Maryes Abbey	34	S Peters church
3	The Inner	35	White Frery
4	Dominican	36	Sherif Street
5	The Bridge	37	The Castle
6	Newmans tour	38	Fish shambles
7	Tranc Castle	39	Woodstown street
8	Wood Key	40	Woodstock lane
9	Marchan Key	41	Price lane
10	The Hospitall	42	Corke streete
11	S Augustines	43	Ormonds Gate
12	The Tollsell	44	Kinshors lane
13	Bridewell	45	S Tony lane
14	Damas stret	46	S Michaels lane
15	Damas Gate	47	Christchurch lane
16	S Andrews church	48	Cocet church
17	S Georges lane	49	Christchurch
18	S Steves stret	50	Schoolhouse lane

51	S Johns church	61	Senni cort lane
52	Castle steret	62	Iohns Heuse
53	Poll Gate	63	S Eames strete
54	S Warbors stret	64	S Iames Gate
55	S warborts church	65	S Cathri church
56	Skinners Rowe	66	S Thomas court
57	S Nicholas church	67	The Come
58	S Nicholas stret	68	Newe Streete
59	S Nicholas Gate	69	S Francis stret
60	The Tolsell	70	S Patricks stret
61	S Michaels church	71	S Patricks church
62	Highe strete	72	S Brides church
63	Back lane	73	S Brides stret
64	New Gate	74	S Schitlebeins
65	S Tomas stret	75	S Keuan stret
66	Newe Rowe	76	Crosse lane
67	The mille	77	Chur on Paule

Enis Kellinq Fort.

Matrona Hibernice.

Plebeia Hibernice.

Hibernice Rustica.

LYMERICKE

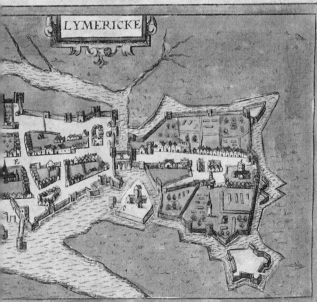

CORCKE

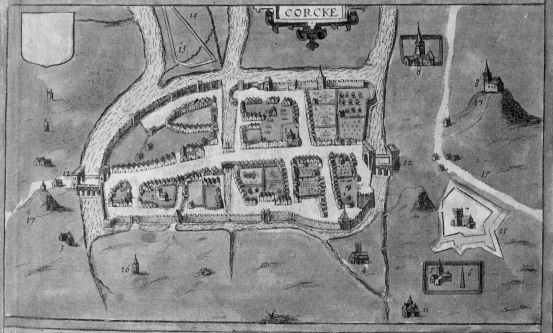

The Key
The Kinges Castle
Thomond bridge
S Frances Abbey
S Peters

L S Dominicks Abbey
N The Bishops house
M S Menshius church
O S Michaells church

1. Chrits Church
2. S Peters Church
3. S Frans Abbey
4. Abbey of ey Isle
5. S Barries Church

6. The Spyre
7. Holly Rode
8. S Steuens Church
9. S Augustines
10. The new Fort

11. The Bishops house
12. The Ports
13. The Entrance Fort
14. The Walke about
15. The Way to Kinsale

16. Shandon Castle
17. The Hills commanding
    the Town

THE RAINBOW BRIDGE IN A SMALL SECTION OF THE QINGMING SCROLL, C.1126 CE Although the date is uncertain and true provenance is unknown, most scholars believe that the famous Qingming scroll was commissioned during the reign of China's emperor Huizong (reigned 1100–1126) and it is conventionally attributed to Zhang Zeduan. The scroll is believed to depict detailed scenes in Kaifeng, the great eastern capital of the Northern Song dynasty of China and one of the most sophisticated urban centres in the world at the time. But there are some who argue it is not a specific place and represents instead a model city – *qingming* refers to a great festival but it also means 'peace and order', the urban ideal. The rainbow bridge is the scroll's key landmark depiction and although Kaifeng did have one, so did several other northern cities. The somewhat unrealistically idyllic representation has even been interpreted as a subtle – highly Chinese – way of implying the opposite. Whatever the truth, the painting captures the urban spirit and animated essence of 12th-century China with its inns, restaurants, busy riverfront, shops, wholesalers, teeming citizenry and travelling traders. **PREVIOUS PAGES**

Cities are places of hopes and dreams, of vision and order, as well as centres for destruction and conflict. Although cities are not creations of the modern era, for many people they represent the core element of life as we live it today, when most of the world's population lives in an urban hub of commerce, technology, transport and social interaction between people, and in communities of often quite diverse cultures. Whereas only a century ago perhaps 10 per cent of humankind lived in a city, now most people do and the world's economic development is characterised by the relentless, and often unrestrained, expansion of our ever-growing urban metropolises.

Globally, cities have become inextricably identified with this sense of progress, success and advancement, whether individual, social or economic; cities are believed to be places where things 'happen'. In fact, historically, this has long been the case, with cities impossible to separate from the evolution of human civilisation.

Trade and religion are two of the oldest practices of humankind, and cities originated and grew to facilitate the complex human webs of exchange involved in both, which have left their marks throughout the millennia on the form and features of our cities – to facilitate the buying and selling of goods and to enable people to gather for matters more transcendant and less material. And just as civilisation grew out of humankind's conscious attempt to control, change and organise our environment, so cartography and mapping arose out of our need for tools to measure, record, understand, navigate, plan and protect our surroundings. Cities – centres of spiritual, economic and political power – became leading centres of mapmaking as well as prime subjects for cartographers.

City maps are among the most popular, as well as oldest, forms of cartographic representation. However, the survival rate of early maps is limited and most maps date only from the last 500 years. Early maps are fragments; sometimes literally in the physical sense, but also because our knowledge of them is incomplete, based upon a partial understanding of the cultural context within which they were made, though it seems quite clear that the world such maps depict was centred upon the culture of origin and that cities loomed large as places that gave meaning and identity to those same cultures.

THE FIRST CIVILISATIONS

The shift to crop cultivation encouraged the production of regular food surpluses, which made it possible for some workers to specialise in other tasks. Urban development rested on agrarian systems that were able to support substantial populations, and these were found first in fertile river valleys, such as the Euphrates and, later, Tigris of Mesopotamia (Iraq), the Nile valley of Egypt, and the Indus valley of modern Pakistan. In Andean America, in what is now Peru, large temple mounds appeared in the central Andes along the Pacific coastal river valleys, in places such as the Supe Valley, from c.2500BCE. In East Asia, the Yellow River was the later basis for Erlitou, founded in about 1900BCE, China's first city.

In Mesopotamia the city-state of Uruk developed in about 3500BCE. The sacred enclosure of a raised mud-brick ziggurat temple complex was an important feature of the early Mesopotamian cities, not only because the priests provided sacral power but also because the temple administered much of the city's land while the priests could record production and store products.

By about 3300BCE walled towns had begun to be built along the Nile in predynastic Egypt, Nekhen, or Hieraconpolis, and Naqada being the earliest. When

TERRACOTTA FRAGMENT, NIPPUR, C.1250BCE
This unique map of the holy city of
Nippur in Mesopotamia dates from the
Kassite period when the city was
reinvigorated by the building of irrigation
canals. Nippur was located on the River
Euphrates, in what is now Iraq, and was
the centre for the worship of the god
Enlil, the most revered deity in the
Sumerian pantheon. The city's first
temple dedicated to Enlil was built for
the ruler of the city-state of Ur in about
2100BCE. The two parallel lines on the left
of the fragment represent the River
Euphrates. Emerging from that, almost at
a right angle at the top, is a canal that
loops down into the city, bisecting the
fragment from top to bottom. Running
parallel to the river and the canal at the
top – and also running across the bottom
– are a moat and Nippur's city walls,
punctuated by six or more gates (denoted
by the clusters of cuneiform). In the
southwest quadrant (bottom left) there is
a park, to the right of which is a ziggurat
temple complex known as the Ekur,
associated with Enlil. On the other side of
the canal is the Ekiur, associated with
Inanna (Ishtar), goddess of fertility. The
temples were the city's dominant
institutions. First settled in around
5000BCE, Nippur was not abandoned until
about 800CE, by which time it was the
seat of a Christian bishop – still a centre
of religion long after Enlil had faded away.
LEFT

the country was unified in about 3100BCE by King
Narmer, the first pharaoh, he founded Memphis as his
capital, which was built on the west bank of the Nile,
south of the delta not far from modern Cairo.

In the Indus Valley, walled settlements were
followed, in about 2500BCE, by major cities, notably
Harappa and Mohenjo-Daro. Spread over 148 acres
(60 hectares), the latter city had a population of maybe
50,000 as well as crucial urban infrastructure in the
form of a sophisticated sewage system.

COMMERCE AND CONFLICT

Because trade was such an important aspect of these
early urban civilisations, long-distance commercial
networks grew by sea and land. These included ports,
such as Byblos (Lebanon), founded c.3100BCE, as well
as Dilmun (Bahrain) and Ras al-Junayz (Oman) to
link to eastern maritime centres, and trading cities
and colonies across the South Asian inland hinterland,
such as Shortughai along the Oxus River in northern
Afghanistan, c.2500BCE.

Competing interests and the need to maintain
control and security encouraged the walling of

settlements in anticipation of large-scale conflict. The
first empire in western Asia was founded in about
2300BCE by Sargon, who united the city-states of
Sumer (southern Mesopotamia) and conquered
neighbouring regions. An empire based on the city
of Ur followed.

Protected by encircling walls and a fortress, Ur was
linked to the Euphrates River by canals, which
provided another inter-urban form of transportation
network for trade. Later came the Babylonian empire
of Hammurabi (reigned 1790–1750BCE). With places
such as Babylon, cities became further associated with
learning, culture, the law and man's management and
modelling of nature.

A Babylonian clay tablet from about 600BCE
provides the earliest known evidence of world
mapping, though the purpose of the map is unclear.
The world map centres on Mesopotamia, with
Babylon shown as an elongated rectangle. Parallel lines
running to and from it represent the River Euphrates.
All these symbols are contained within a circle that
represents the ocean. If this map can be interpreted as
revealing a sense of cultural self-confidence, perhaps it

THE LOCATION OF CITIES IN NORTHERN IRAQ, 11TH CENTURY This map marks the Dejlah and Forat (Tigris and Euphrates) rivers as well as important towns and cities located along each, including Baghdad, Mousul, Racca, Samisat, Amid, Beled and others over several days distance within the region known as al-Jazirah (approximately northern Iraq). The map, executed in the style known as the Balkhi School, is from the *Kitab al-masalik wa al-mamalik*, or *Book of Routes and Provinces*, an 11th-century Arabic geography composed in the tenth century by Abu Ishaq Ibrahim al-Istakhri. The cardinal points are written in angular kufic gold script. By 1055, with the fall of Baghdad, this area had come under the control of the Seljuk Turks, originally from Central Asia. In 1071, the Seljuks defeated the Byzantine emperor Romanus Diogenes, and in 1086 the Seljuks took Amid from the Kurdish Marwanid dynasty, which ruled the area. By the 12th century the Seljuks had reunited all of the old Abbasid territories. The *Kitab* originated with the government postal service in Baghdad to describe the overland and maritime routes that linked the Abbasid realm to the world, including a description of the sea routes to the great maritime cities of India, Malaya, Indonesia and China. **RIGHT**

is no coincidence that cities also became the focus of empires engaging in territorial conquest, helping to expand the influence of urban civilisation and offering up a template for successful replication.

The stone reliefs from the palace of Nineveh, the capital of the Assyrian Empire (c.950–612BCE), depict sieges of cities. In turn, the Babylonian Empire under Nebuchadnezzar II extended to Palestine, where Jerusalem was destroyed in 587BCE, only to be overthrown in turn by the Persians in 539BCE.

CITIES – A GLOBAL PHENOMENON

In ancient China, then as now the world's most populous country, a strong economy, built upon the production of millet and rice, combined with a sophisticated administrative system that meant the state was able to support a large urban population. Under the Shang Dynasty (c.1800–1027BCE), there were a number of capital cities, notably Erlitou and Anyang. Its Zhou successor (1027–403BCE) again had a number of capitals: it is from the Zhou Dynasty that we have the first documented city planning. The principles of Zhou urban design, which continued to underpin Chinese grid layouts into the modern era, were based upon a holy square system derived from a mixture of cosmology, astrology, geomancy and numerology.

During the Qin Dynasty (221–206BCE), there was a series of administrative centres under the imperial capital of Xianyang. This was also the case in the Han era (206BCE–220CE) with its successive capitals of Chang'an and Luoyang, as well as its thriving coastal maritime cities such as Fuzhou. Later, under the Tang (618–970), Chang'an (modern-day Xi'an) was the capital and it had a population (within and outside the walls) of about two million by the eighth century. The city's symmetrical layout was used to organise specialised and orderly functional neighbourhoods, the demarcation of which arose out of by then deeply-rooted Chinese ideas about the spiritual efficacy of spatial arrangements and alignments – ideas that were diffused to various degrees throughout East Asia. Ancient China's urbanisation was such that in Tang-dynasty China there were more than ten cities with populations of 300,000-plus. During the later Song dynasty (960–1279), the merchant's entrepot and metropolis of Hangzhou had a million residents at a time when London had around 15,000. The commercial wealth of 11th-century Kaifeng, a

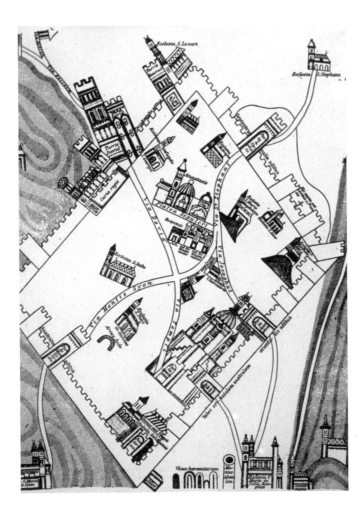

canalised capital of the Northern Song in north-central China so beautifully depicted in the Qingming scroll by Zhang Zeduan, far outstripped that of any European city at the time.

Well before Europeans settled there, cities had also developed in the New World of the Americas, notably the hilltop Zapotec city of Monte Alban in central Oxaca (southern Mexico) in about 500BCE and El Mirador, the largest early Maya city by about 250BCE. To the west, in central Mexico, Teotihuacan, a grid city with temple-topped pyramids, had 125,000–200,000 inhabitants by 500CE. In South America, Tiwanaku (Tiahuanaco) on the shore of Lake Titicaca in modern-day Bolivia, a centre of religious activity, had up to 40,000 inhabitants.

In the case of another city-based empire – Rome – the purpose of the city was clear and highly important: the display of power. A large-scale plan of the city, the *Forma Urbis Romae*, was incised on a wall for public view. The display of maps was used by Julius Caesar and other leaders to demonstrate how Rome was fulfilling its destiny through imperial expansion.

The rise of Rome stimulated an interest in the wider world among the polity's leadership, which resulted in the earliest known globe of the Earth being produced there in about 150BCE by the Greek scholar Crates of Mallos.

A city with a large population had to be sustained through well-organised, well-maintained infrastructure, and this created a need for maps that could record useful information in graphic form. Rome's population may have reached about a million in the second century CE. The supply of goods to support this population was a major economic, governmental and logistical achievement, notably the supply of grain from Sicily, Tunisia and Egypt, with Alexandria operating as a key entrepot. Major warehouses in the southwest of Rome along the River Tiber testify to the importance of trade. Rome also depended on a network of aqueducts to supply it with water.

Roman civilisation was based on an urban culture and forms of organisation. Established specifically as military bases and centres for trade and government, cities developed across the Roman world. From Cologne to York, numerous cities began in this way. Under the later Roman Empire, Rome's position as imperial capital was in part delegated to other cities, notably Constantinople, Milan and Trier.

While Rome provided a stage for cartographic display, the major advances in knowledge were made elsewhere. Eratosthenes (c.276–194BCE), a Greek astronomer who became the chief librarian at Alexandria, in Egypt, calculated the Earth's circumference with great accuracy. Also in Alexandria, by then part of the Roman Empire and a major centre of intellectual life, Ptolemy (Claudius Ptolemaeus, c.90–c.168CE) drew up a world gazetteer that included an estimate of geographical coordinates.

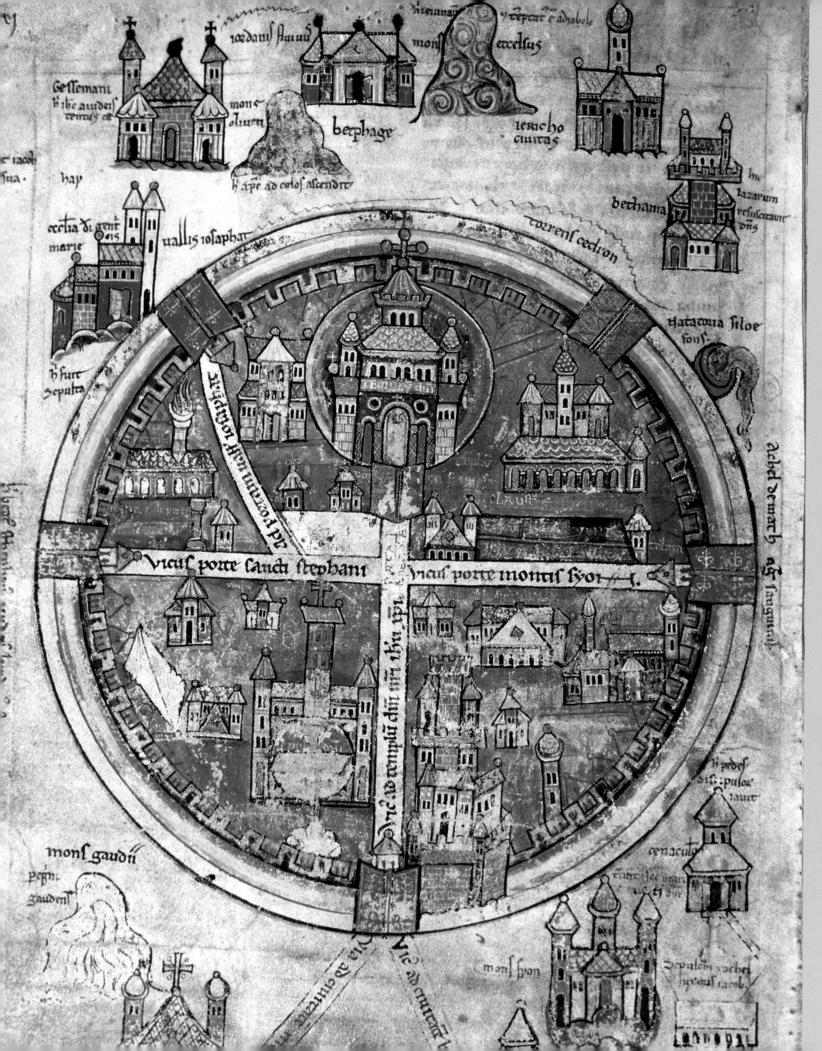

Introduction

MEDIEVAL MAP OF JERUSALEM This map comes
from Robert le Moine de Reims, French
abbot of St Remy, who was present at the
conquest of Jerusalem in 1099 and wrote
Chronicles of the Crusades. Although
plans of Jerusalem and the Holy Places
had accompanied the account of Arculf's
pilgrimage as early as 670, a regular
sequence of maps of the city only began
following its capture in the First Crusade
(1096–1099). Medieval maps usually
transmitted quite basic information
within the framework of a spatial
representation – they were not
attempting to represent geographical
reality so much as convey limited
information to a defined group. For
example, a pilgrim map would indicate
shrines and hostels. Most of the maps of
Jerusalem were stylised, giving the city a
circular, diagrammatic wall in which the
thoroughfares comprise a cross. Thus,
religious symbolism took precedence
over accuracy. The map was illustrated
with key religious sites, at many of which
churches stood. Robert le Moine's map is
not a typical T–O map or *mappa mundi*
of the period, an imaging of the world in
geographic–theological terms, but the
Christian symbolism remains clear. As well
as churches, there appear to be temples
and mosques. **LEFT**

Metropolis

LONDON TO BEAUVAIS SECTION OF AN ITINERARY, FROM *HISTORIA ANGLORUM*, (C.1200–59)

Matthew Paris, a monk in the Benedictine abbey of St Albans, was one of the finest mapmakers of his day and an important chronicler. At a time when book production was quite a communal affair, Paris was unusual in being a compiler, a scribe, a composer of original material and an illuminator. Having entered the monastery in 1217, Paris produced his work during the reign of King Henry III (r.1216–1272), which was a period of economic expansion that saw a significant increase in town foundations and growth. In his *Historia Anglorum*, Paris featured a detailed itinerary from London to Jerusalem via France and Apulia, of which this folio is from the first leg. Each day of the journey is charted in linear format, with prominent features en route depicted. This London–Beauvais section is undertaken via Rochester, Canterbury and Dover, then Wissant, Boulogne, Montreuil, St Valery sur Somme, Abbeville, St Riquier and Poix de Picardie to Beauvais. Some scholars argue that as accurate as such an itinerary might be, the real journey was spiritual and was made in the imagination of would-be pilgrims – in Paris's immediate surroundings, those in the monastic communities. The map of Jerusalem itself appears, perhaps for theological reasons, on the seventh page. RIGHT

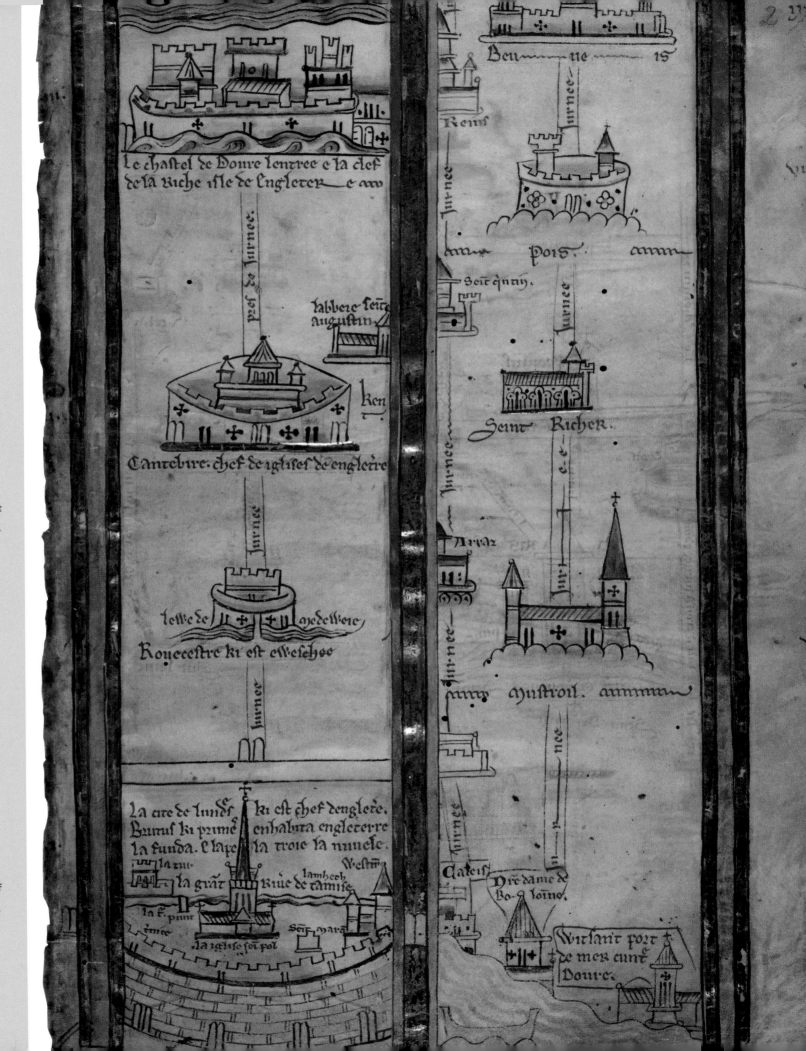

In the *Peutinger Map*, made between 335 and 366, but which only survives as a 12th-century copy, cities were used as the means to organise space. The map shows main roads and depicts the principal cities, using different, stylised pictures for each one, a device still in use many centuries later. In a similar vein, the *Ravenna Cosmography* is a list of more than 5,000 place names covering the Roman Empire, which was drawn up in about 700CE by an anonymous cleric at Ravenna, the last capital of the Western Empire. It has been suggested that the compiler had access to a range of official maps.

Technology-based modern archaeological research now provides us with far more information than is available from surviving Classical maps, notably for Rome. Maps can only cover so much, but the creation of any map also involves a decision about what to omit from it, which means that the selective depiction of history is an issue for those studying maps of the past. In 2010, I noted that the impressive Musée Gallo-Romain in the French city of Vienne, an important Roman site in the Rhone Valley, did not acknowledge the existence of slavery. Thus, the site includes the remains of a *fullonica* (fulling mill, c.2CE) without mentioning the labour force. This serves as a reminder about the possible importance of what maps leave out.

CENTRES OF IMPORTANCE

There are few maps depicting cities from the so-called post-Roman Dark Ages, although a dearth of maps should not lead to the conclusion that there were few places of significance. In the *Life of Charlemagne*, written between c.829 and 836, Einhard stated that the emperor had plans of Constantinople and Rome, the centres of imperial resonance, which he sought to recreate in his capital Aachen, engraved on tables of gold and silver. These plans, however, did not survive.

In many places, historic urban life has left only a limited impact in the cartographic record. This was the case in Africa where trade, politics and religion were also important to the development of cities. Capitals of states with important commercial positions, such as Axum (northern Ethiopia) thrived accordingly – in its case from c.100BCE to c.600CE. Subsequently, in the sahel belt of West Africa and along the coast of East Africa, the spread of Islam and the growth of trade were linked to the expansion of cities, such as Jenne, Timbuktu and Gao on the River Niger, Kano in northern Nigeria, and Mogadishu, Malindi, Mombasa, Kilwa and Sofala on the Indian Ocean. There were also cities in parts of the African interior, such as Great Zimbabwe from the seventh century and Gondar in Ethiopia from the 1630s. In the early 14th century the emperor of Mali, Mansa Musa, undertook a famous pilgrimage to Mecca in a journey that revealed to the world the wealth of parts of West Africa. His trans-Sahara caravan to Cairo brought Mali to the attention of cartographers in Spain, Germany and Italy, as well as piquing the interest of acquisitive rulers in North Africa, and he was famously depicted in a contemporary map dating to the 1330s.

One city for which many depictions have survived, however, is Jerusalem, a historic centre of Christianity, a place of holy pilgrimage from as far afield as the British Isles and a powerful symbol in Christian thought. The Bible was a significant inspiration for cartography in medieval Europe. Not only was there widespread interest in the location of places mentioned in it, but, in addition, *mappae mundi* inscribed the Biblical story as a central theme in the depiction of the world. These maps typically employed a tripartite internal division with a T–O shape depicting three continents – Asia, Europe, Africa – all contained

Introduction

THE PORTOLAN CHART KNOWN AS THE CATALAN ATLAS, 1375 Until the era of European colonisation, trans- and sub-Saharan Africa had little presence in maps although trade routes had been established across the continent for centuries. Islam's spread out of Arabia gave further impetus to the growth of cities in West Africa. Once the commercial links between the sahel belt and the Indian Ocean and Arabia were in place, and Islam had taken root, those routes became the ways along which pilgrims could undertake the Hajj to Mecca. Catalan chartmakers from the mid-14th century began to include references to this, and a collection of other information (such as natural resources), in their maps, which were actually types of navigator's portolan chart (hence the geometric rhumb/nodal lines everywhere). For example, Mecca or other important Muslim-controlled cities might be identified by a banner bearing a crescent. One of the most famous rulers of the age was Mali's emperor Mansa Musa, who controlled rich goldfields from Gambia and Senegal to Gao on the River Niger. Cartographer Abraham Cresques, commissioned by Pedro IV of Aragon, has depicted Mali's ruler with a gold crown, ingot and sceptre, topped with a gold fleur-de-lis. The great centre of learning at the eastern end of this long route was Timbuktu. Mansa Musa's Hajj in 1324 had involved such extravagant expenditure in Cairo that it was famous in Europe. **LEFT**

A MAP OF SICILY, C.1220–1320 This map is from a copy of an illustrated Islamic cosmography, compiled by an unknown author in the 11th century, known as *The Book of Curiosities*. Drawing upon the *Geography* of Ptolemy, its focus is on Islamic commercial centres in the Mediterranean. The map of Sicily, then an Arab-ruled territory under pressure from Christian Norman assailants, is dominated by its leading city, Palermo, and its hinterland. The Old City of Palermo is represented as a circular enclosure in red, broken by ten named gates. Immediately above is an indentation representing the harbour, outside the walls; to either side of it are towers. On the eastern side of the harbour, the arsenal is shown. The ruler's palace (identifiable by the onion dome) lies a little further east. Various quarters of the city are annotated, including the walled 'Quarter of the Europeans' as well as the new quarter of al-Ja'farÐya. Places where water sources originate are shown outside the walls. The suburbs of the city are presented as spreading over most of the island, which suggests that Palermo was not adjusted proportionally when a copyist reduced the size of the island. The net effect is to have the island dominated by the city.
RIGHT

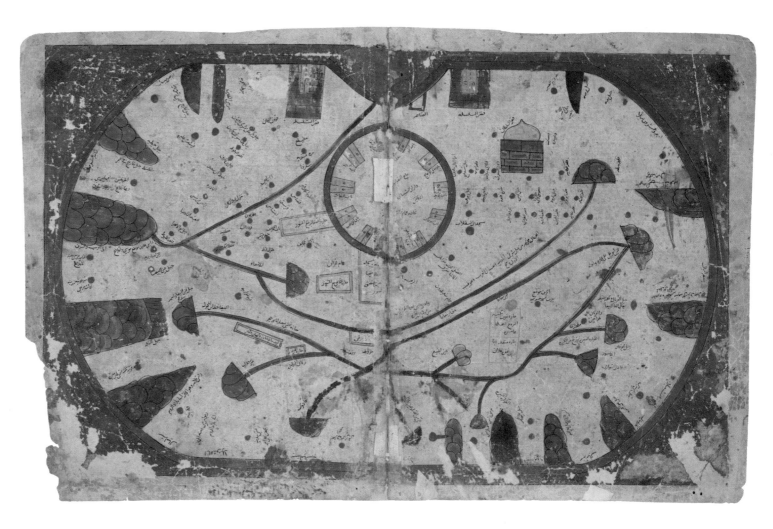

within a circle, the O, with the horizontal bar of the T representing the Don and Nile rivers separating Asia from the other two continents. The body of the T itself was the Mediterranean Sea and a symbol of the Christian cross – and Jerusalem was positioned at the centre of the world, not so much in a literal sense as a spiritual one.

The role of Jerusalem was further underlined by the Crusades that began in the 1090s, to recover for Christendom the sites associated with the stories in the Bible and the death of Jesus. Although plans of Jerusalem and the Holy Places had accompanied the account of Arculf's pilgrimage as early as 670, a regular sequence of maps of the city did not begin until after its capture in the First Crusade (1096–1099), as with the map of the city from the *Chronicles of the Crusades* by Robert le Moine de Reims, French abbot of St Remy, who was present at the conquest of Jerusalem. Most of these maps were stylised, giving the works a circular, diagrammatic wall in which the city's thoroughfares comprise a cross. Thus, symbolism took precedence over accuracy.

Once Jerusalem had been lost anew to Islam in 1187, the city was depicted as the necessary goal for Christian Europe. The early 14th-century *Liber Secretorum Fidelium Crucis*, or *The Book of The Secrets of the Faithful of the Cross*, by the Venetian historian Marino Sanudo, presented to Pope John XXII in 1321, was a call for a crusade to regain the Holy Land. This work, which belongs to a genre known as 'recovery literature', includes maps by Genoese cartographer Pietro Vesconte that depict the former Crusader cities of Jerusalem and Acre.

Cities were largely rendered pictorially in these early maps, as with the depiction of London by the St Albans' monk Matthew Paris in the itinerary from London to Jerusalem that prefaces his *Chronica Majora* of c.1252. Few pilgrims will actually have carried such a map on the journey, using it instead at home as a devotional tool for pious contemplation.

DETAIL OF CATHAY, FROM FRA MAURO'S *MAPPA MUNDI*, 1448–1453 Fra Mauro was a monk at the monastery of San Michele on Murano island, near Venice, and in the 1450s he produced perhaps the first world map. Believed to have been commissioned by King Afonso V of Portugal, the map created by Fra Mauro offered a descriptive cartography rather than a mathematical one using projection, though Mauro tried to use scientific methods – for example, referring to Portuguese nautical charts and accounts from and interviews with travellers – rather than simply the supposed authority of some Classical sources. Fra Mauro's emphasis was on pictures and in Venice he was well placed to draw upon sources with knowledge of the Orient. That about East Asia was amost certainly gleaned from Marco Polo, who was also a source for the *Catalan Atlas*. Cathay was the Anglicised name given to Catai, or northern China, which was presented by drawing upon traditional European-style imagery (albeit that north is at the bottom). It is thought that the Doge's palace in Venice contained a mural of Polo's travels that was later destroyed by fire and it may also have provided Mauro with a visual source. **LEFT**

Significantly, the traveller starts from London where pilgrims could visit the shrine of Edward the Confessor in Westminster, which is shown, together with the White Tower of the Tower of London and one of the earliest views of the great medieval cathedral of St Paul's, with its tall spire. At the time of its depiction the cathedral's prominent vertical feature was the spire; while admittedly only timber, it reached 520 feet (158 metres), which is much higher than Salisbury's at 404 feet (123 metres). This spire – a feature only surpassed in London by the Post Office

Tower of 1964 – was a prominent feature in all views of the city, which suggests both that it was distinctive and that the citizens took pride in it. In 1561 the spire was destroyed by lightning. Paris also shows the city's castellated walls and lists some of the city gates.

Although in the case of Jerusalem control over one of Christendom's most sacred places had been lost, throughout Europe there were many other cities that were important for political and economic reasons. Textile manufacturing centres were particularly significant in medieval Europe. Places such as Milan,

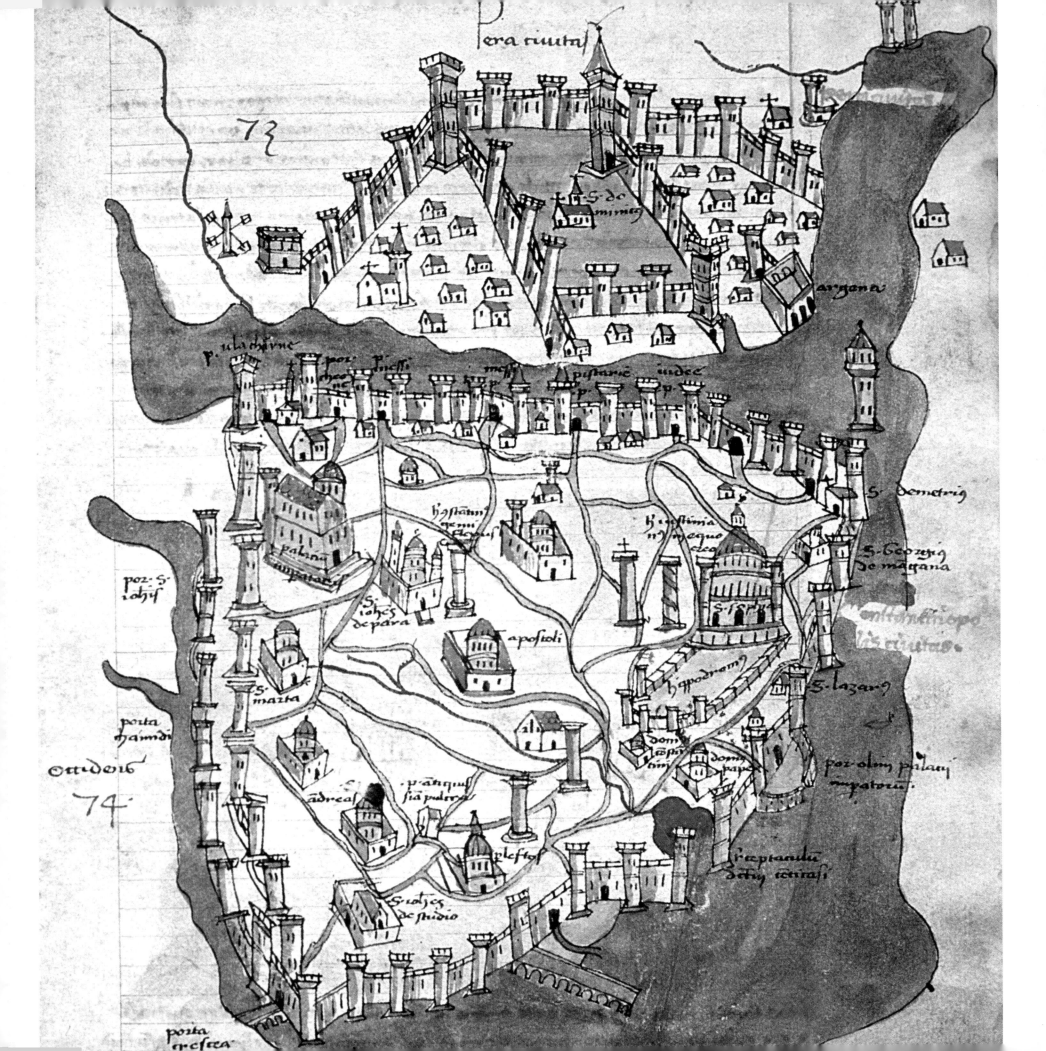

Cremona, Ghent and Bruges were central to a thriving commercial expansion that encouraged new urban settlements and growth, notably around the Baltic and Mediterranean, with major cities such as Lübeck, Riga and Venice. The Hanseatic League of Baltic cities represented a federation of urban power, while Venice, like Genoa, ruled a maritime-based territorial empire that was similar to Classical Athens. The most populous city in Latin Christendom was Paris, which in 1300 had a population of maybe 200,000. As well as being the capital of France, the most powerful state in western Europe, Paris was a major intellectual centre thanks to the reputation of its university. In Eastern Christendom that honour belonged to Constantinople, which may have had a population of as much as half a million by the fifth century, long before Florentine traveller Cristiforo Buondelmonti produced his schematic bird's-eye view of the somewhat reduced imperial city (following its traumatic capture in 1204 during the Fourth Crusade) in the early 15th century. Indeed, the only city in Europe to rival Constantinople in size lay not in the Christian West but in Iberia: Muslim Córdoba.

TECHNOLOGY AND MEANING

In the scientifically advanced Islamic world, a separate strand of cartography could be found, albeit one that drew on the Greco-Roman tradition. By the ninth century, Arab geographers and their patrons had made cities such as Baghdad, Cairo and Damascus into major centres of cartography. The importance of cities in the Islamic world rested on political power and commercial significance. Baghdad was the capital of the Abbasid caliphate (762–1258; although Samarra further up the Tigris had that role from 836 to 892); Cairo of the Tulunids (868–905) and Fatimids (909–1171); and Damascus of the Umayyad caliphate (661–

750), with Córdoba in southern Spain constituting a breakaway Umayyad state (756–1031). Fez in North Africa was established at the end of the eighth century because Idriss I decided that Volubilis, a city founded by Carthaginian traders in the third century BCE and once a major Roman settlement, was too small and he wanted a new capital. The expansion of Islam from its heartland in the Arabian Peninsula and its transmission eastwards, both southwards towards South Asia and northwards along the Silk Road through Central Asia, led subsequently to cities such as Isfahan, Delhi and Samarkand becoming important as the capitals of Islamic states.

As in the medieval West, Islamic mapping was diverse and it included world maps, centred on the sacred city of Mecca and a destination for many pilgrims, as well as bird's-eye views of important spiritual and mercantile cities. These maps were often laden with symbolic meanings, rather than capturing details of shape and topography, although Arab cartographers did use carefully calculated mathematical data with which they were able to determine more accurate map projections than their European counterparts.

The loss of territories in the Eastern Roman Empire (Byzantium), notably Egypt and Syria, to conquest and gradual cultural consolidation by Islam from the seventh century onwards, meant the loss to Christendom of Classical knowledge in the shape of Greek geographic information and ideas. The Islamic world was able to draw not only on Ptolemy and the Greek development of Babylonian astral religion in order to provide astronomical and celestial know-how, but also to expand its cartographic knowledge through information, ideas and methods flooding back from far-flung lands through conquests, trade and travel. Caravan routes linked the Orient to the Middle East,

and also crossed the Sahara Desert to West Africa, while Arab traders, benefiting from their astronomical knowledge and employing a star compass, sailed the Indian Ocean and the Mediterranean. Making use of monsoon winds, Arab mariners and merchants sailed eastwards in the Indian Ocean, and, in the late eighth century, began trading with East Asian commercial hubs such as Guangzhou (Canton).

The importance of maritime trade to our ocean-girdled planet ensured that port cities were found across much of the world. This was especially the case when ports could service trading routes between major economies, as with Aden, Kulam Mali in southern India and Kataha at the northern end of the Strait of Malacca in about 1000 and Aden, Malacca and Brunei half a millennium later. When Europeans initiated the great age of oceanic exploration that did so much to begin what has been called the 'triumph of the West', sailors and explorers were to provide the cartography industry with a significant boost.

European maps – including local area maps; route maps, such as the British Gough map; and the nautical charts known as portolan charts – were often embellished with drawings of significant buildings. This was a pictogram-like element, not unlike the images used on seals and coins, and accurate topographical detail was not yet a factor because of imprecise knowledge of how best to measure such

features, as well as technological limitations over the means to represent them. Symbolism therefore plays a part in European maps, just as it does in sometimes more complex forms (to our way of understanding) in other cultures – for example, the indigenous peoples of the Americas.

PLACES OF TRANSFORMATION

Prior to the age of print, which began in Europe in the 15th century (and several centuries earlier in China), maps were produced by a number of means, some quite ephemeral, but for the most part they required painting by hand. Engraving and then lithographic print technology meant that maps acquired greater permanence, or at least durability, because they could be more easily and accurately copied and distributed at less expense than previously; colour printing enabled maps to become far more sophisticated; and the emergence of more universally understood forms of quantification and representation meant that maps became more widely comprehensible. Prior to such developments, maps – their projections, and even the record of time and place that they represent – can often only be understood by viewing them carefully through a prism of cultural awareness, as is well demonstrated by comparing Tenochtitlán from two different perspectives.

As well as providing major subjects and topics for mapping, and challenges for effective depiction, cities

represented to individuals places of opportunities for betterment, albeit perhaps not equality of opportunity. The sustained migration to cities that has characterised much of the population history of the last two millennia is testimony to this interpretation. Each migrant represented an individual decision that life might be improved in a city. For many, this proved illusory and rural penury was translated into urban poverty. Nevertheless, social control was laxer in cities, because it was harder to enforce. Urban populations were not necessarily radical in their politics or beliefs, but urban life provided the forcing ground for most new ideas, both élite and popular, and also offered people new and transforming experiences.

This book discusses these themes and, in so doing, throws significant light both on the history of the city and on that of mapping. The subject of the city in maps is important and interesting because it ranges so widely, offering a leading account of the environment in which much of the world's population lived, especially that portion of humankind that has most shaped things politically, economically, intellectually and culturally.

While some city maps might have begun merely as aids for visiting strangers, countless others deal with far more complex and fascinating issues than how to get from A to Z. These maps provide us with an immensely rich historical archive that traces urban growth and spatial use, including the deliberate human redesignation of space. A map can not only help the curious to comprehend the world around them, it can be a dynamic agent to set out and shape competing visions of urban development; it can record the multilayered human activities which take place within these urban spaces, even demonstrating how those activities can be significantly shaped by that same urban environment. In other words, maps of our cities offer us far more than a bird's-eye view of streetscapes, endlessly fascinating though those undoubtedly are, because through maps-as-information-graphics we can view historical and modern patterns of poverty, crime, disease, ethnic settlement and countless more data.

Metropolis is organised into five chronological chapters that trace the city in maps produced in all their variety during the last half a millennium – from Asia to the Americas, during the flat-map centuries of times gone by through to an era of GPS-derived imagery and digitised, three-dimensional computerised and interactive mapping that was previously impossible to produce. The book concludes with a survey of how the city-to-be has been envisaged in previous eras and how we are now planning and building what are expected to be the ecologically sustainable cities of the future, while trying to retain the attractions of what has drawn so many of us to our cities in the past.

TENOCHTITLÁN: Causeway city

Native documentary sources date to 1325 the foundation by the Mexica (Aztec) people of a lagoon city of canals and causeways on a marshy island in Lake Texcoco. However, the pictorial representation of this event is a blend of reality and myth, requiring specialist knowledge to interpret. The only surviving indigenous map of the city is in a codex of about 1541, created for Antonio de Mendoza, the Spanish viceroy from 1535 to 1550. Unlike European maps, that drawn by the Aztecs reflects time instead of space, emphasising the degree to which maps can incorporate culturally specific information, perhaps encoded with symbolism that is unfathomable to outsiders.

TENOCHTITLÁN IN THE *CODEX MENDOZA*, 1541 This map, produced by indigenous scribes known as *tlacuilos*, depicts the founding of Tenochtitlán. Rather than offering an accurate pictorial representation, the abstract map uses symbols which tell the city's invented past and its social layout. This contrasted with the European mapping of the period, which was more to do with the use of space. The map is dominated by an eagle (symbol of the sun) on a prickly pear cactus growing on a stone, which refers to the city's origin myth. The eagle is associated with Huizilopochtli, the god of war and patron deity of the Aztecs, who had told the leaders to look for this sign of where to settle. The dark-skinned leader (left of the cactus) is the foremost priest – the only male with a speech bubble. Beneath the stone glyph are arrows and a shield, which means war. The shield design signifies Tenochtitlán. The diagonal cross is a stylised plan to depict an island quartered by two waterways. The quadrants delineate the social organisation of the city into four wards. The structure at the top may be the humble origin of the Templo Mayor. The skull rack (*tzoompantli*) attests to the practice of human sacrifice. **LEFT**

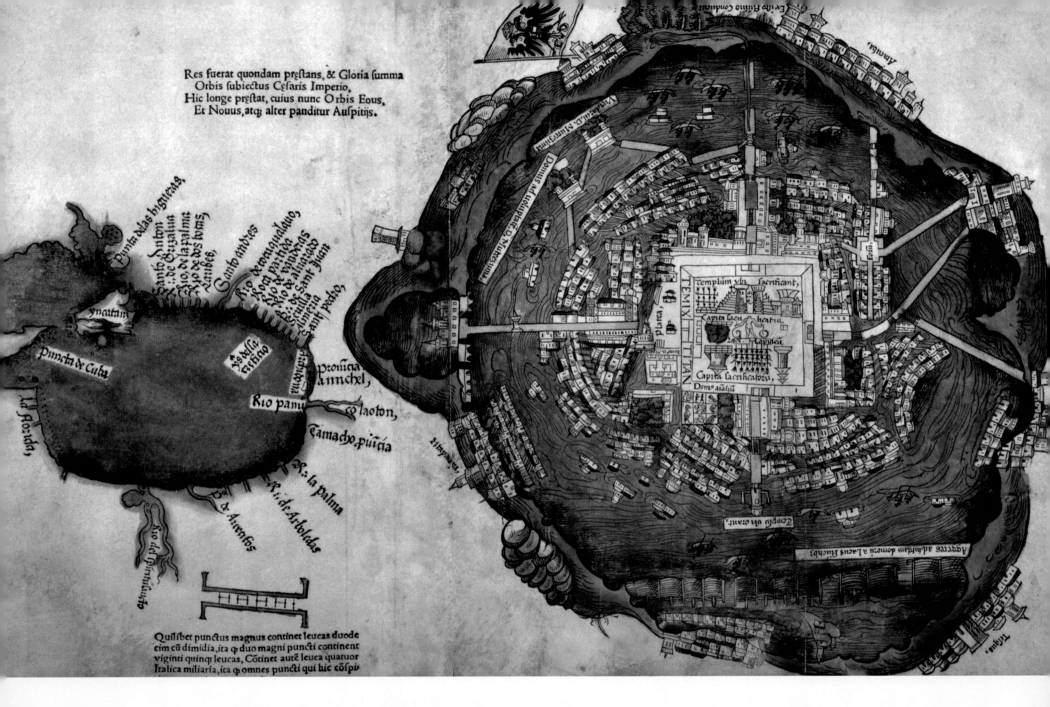

Res fuerat quondam prestans, & Gloria summa
Orbis subiectus Cesaris Imperio,
Hic longe prestat, cuius nunc Orbis Eous,
Et Nouus, atq; alter panditur Auspitijs.

Quilibet punctus magnus continet leucas duode
cim cū dimidia, ita q̄ duo magni puncti continent
viginti quinq; leucas, Cōtinet autē leuca quatuor
Italica miliaria, ita q̄ omnes puncti qui hic cōspi

TENOCHTITLÁN, C. 1524 This woodcut plan of the great island city,
alongside a Gulf of Mexico inset, was produced for a letter sent
by Hernán Cortés to Emperor Charles V (one of five he sent
between 1519 and 1526). The map fuses European and
indigenous knowledge and spatial understandings. Although the
Valley of Mexico setting is conventionalised, the arrangement of
the low-roofed city is well conveyed, with causeways and
nearby towns. At the bottom, the dyke that was the source of
the city's fresh water is shown. The heart of the city is the Great
Plaza, where the stepped pyramid (correctly aligned here) was
yet to be replaced by a huge cathedral. As La Plaza de la
Constitución, this remains the great public space of Mexico
City. The Spanish pulled down the centre of the city, but the
street layout stayed virtually intact. ABOVE

Metropolis

THE IDEAL CITY, BY PIERO DELLA FRANCESCA, LATE 15TH CENTURY During the Renaissance the wealthy and powerful became great patrons of art and architecture. The duke of Urbino commissioned an architect to design a palace complex that was a model city and it was this aspiration of planned perfection that della Francesca sought to capture in his fresco in Urbino's Palazzo Ducale. His imagined urban utopia (a three-panel work known as *The Ideal City*), has streets lined with elegant Renaissance palaces flanking a Roman-inspired rotunda church at the centre, which has been mathematically constructed to contrive perspective via converging lines. **PREVIOUS PAGES**

LISBON, 1598, VOLUME V OF *CIVITATES ORBIS TERRARUM* BY BRAUN AND HOGENBERG This view of Lisbon is an important record of the great maritime city at the height of its influence in the Age of Exploration. More than two-thirds was destroyed by an earthquake in the mid-18th century. The city plan is accompanied by a key identifying 140 features. Dominating the skyline in the centre is the ruined fortress of Castelo Sao Jorge, which fell to the Moors in the eighth century and was only recovered by Afonso I in 1147. Near the riverfront is the oldest part of the city, Baixa, with its medieval street pattern, which gives way to the Bairro Alto, or upper town. In the foreground, perpendicular to the river is the Ribeira palace, which was lost in the earthquake. **OPPOSITE**

During this period, the world's major cities were situated in Asia, where two-thirds of the world's population lived. Moreover, there were significant changes in the urban hierarchy in Asia, and both economic and political factors played a role. In north India, Agra was founded in 1505 as the capital of the Lodi sultanate, and it therefore began its development into a major centre.

THE WEST EMERGES

However, it was the European voyages of exploration that resulted in European empires in the Americas and a direct trade route to India which ensured that it was the cities of these expanding Europe-based maritime empires that were of growing global importance. These voyages also created a need for new maps as the Western 'known world' was dramatically expanded.

Initially, Lisbon and Seville were the two key cities that facilitated Portuguese and Spanish expansion. Established within the royal palace on the riverbank at Lisbon, the *Casa da India* and the *Casa da Guiné e da Mina* were part of a state monopoly run by the same director and had several treasuries and administrators. These offices supervised the loading and unloading, and dealt with all commercial matters relating to contracts, dues, and so on. The *Armazém de Guine India* (Storehouse of Guinea and the Indies) dealt with nautical issues, from the dockyards to the supply of marine charts.

In Seville, Spain's commerce with the Americas was controlled similarly from the *Casa de Contratación*, or *Casa de las Indias*, founded in 1503. A cartographic department was established within it in 1508. (In 1717 the Casa was moved from Seville to Cádiz.)

The far-flung empires and trading systems of Portugal and Spain transformed urban networks. Existing cities beyond Europe were seized as bases –

notably Calicut, Goa (1510) and Malacca (1511) and by Portugal – and new ones were founded in order to control existing or forge new routes. Thus, to take advantage of the deep water of the largest natural port in the Caribbean, Spain established Havana in 1519. An earlier attempt in 1515 at nearby Batabanó had been abandoned because of infested swampland and the lack of a sheltered harbour. In 1533, the governor moved his capital from Santiago de Cuba to Havana.

The process of colonial expansion was cumulative. Thus, Panama, founded in 1519, became the base for Francisco Pizarro's expeditions to Peru in 1524, 1526 and 1531–1532. The new cities were not only governmental and military centres, but also places from which Christianity could be preached, as with Manila (part of the Archdiocese of Mexico until the late 1570s) in the Philippines, where the cathedral was built in 1581. The cities built by the Russians as they expanded their control across Siberia had a similar character.

ENTERPRISE AND TECHNOLOGY

The mapping of cities in the 16th century was dominated by Western cartographers. This reflected the role played in Western mapmaking by entrepreneurial capitalism, which was not seen elsewhere in the world, and also the opportunities that were provided for this enterprise by the spread of printing technology. Although maps had been printed earlier, the first printing of maps in Europe occurred in the 1470s and, thereafter, map production spread rapidly – a proliferation that was aided by a lack of centralised control, in contrast to the situation prevailing in Ming dynasty China. Indeed, it was independent or autonomous cities that dominated map production, notably Venice, Antwerp and Frankfurt. It was scarcely surprising, therefore, that cities were one

of the main topics for mapping, with a proper pride of place playing a major role.

In the six-volume *Civitates Orbis Terrarum*, produced between 1572 and 1617 by Georg Braun, the editor, and Frans Hogenberg, the engraver, a comprehensive selection of 546 bird's-eye views and map views of cities – mainly European – was presented: a unique view of urban life at the turn of the 16th century. The first atlas of towns, in fact. Texts appeared on the obverse of the plates, with image and information complementing one another. Braun and Hogenberg's work is widely regarded as one of the greatest achievements in European cartography of the age. The pair had dozens of contributors, most of whom relied on existing maps, though none were as important as the Antwerp-based Flemish painter Georg (Joris) Hoefnagel, who had produced many drawings of cities and towns during widespread travels around Europe. Some of the views, such as that of Venice, have what was considered at the time to be a national security

Metropolis

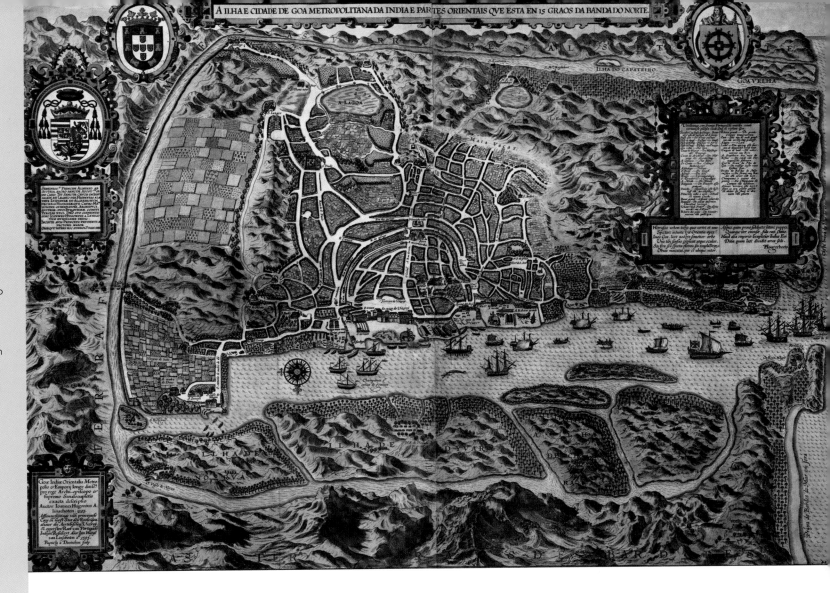

MAP OF GOA, INDIA, BY JOHANNES BAPTISTA VAN DOETECHUM THE YOUNGER, 1595 Van Doetechum was a member of an Antwerp family of cartographers who migrated north to beyond the reach of Spanish military power, first to Deventer and then to Haarlem. The family had a distinctive style of engraving and etching, and made a significant contribution to many books about navigation and exploration during this major era of the voyages of discovery. In 1510 the Portuguese had captured Goa, which had been a key staging post in trade – especially of spices – in the Arabian Sea region for more than a millennium, going back to the Greeks and Romans. The Portuguese made Goa their capital for their Indian Ocean posssessions, the Estado da India, and it also became an important base for Christian missionaries, notably Francis Xavier, from 1542 onwards. Van Doetechum's map shows the natural harbour set in protective hills and with the Mandovi River flowing to the sea. As well as spices, fine muslins and cotton, pearls and diamonds, were shipped through Goa. **RIGHT**

device in the form of people wearing local dress, added by Braun in the belief that Islam forbade the Ottoman Turks – a power feared throughout Europe, which might want to study the plans for military purposes – from looking at the maps because they featured human representations. Braun was supported by Abraham Ortelius, whose earlier *Theatrum Orbis Terrarum* (1570) was one of the first true atlases. Braun and Hogenberg meant their work to be a companion volume aimed at a more popular market, purely because of the appeal of the pictures of cities.

Technological change also led to more accurate maps, which in turn led to greater expectations from urban mapping. Initially, printing used woodblocks, which ensured that maps could be more speedily created and more widely distributed than was possible with manuscript maps. However, from the mid-16th century, woodblocks gave way to engraved copper plates, and in place of the screw press came the rolling

press, using the copperplates to deliver faster production and greater uniformity. Because copper plates were easier to revise than woodblocks, the change contributed to a growing appetite for novelty and precision.

The transformation in production methods was linked to developments in depiction. Earlier European maps were generally picture or itinerary maps, and were not drawn to a consistent scale. However, in the 16th century, drawing to scale became more significant. This was related to new techniques in drafting and presentation. The increasing use of the compass in surveying and mapmaking in Europe from the late 15th century was reflected in a growing tendency to draw local maps with north at the top.

Triangulation was introduced, the plane table and the theodolite developed, and a more consistent and sophisticated use of uniform conventional symbols in place of pictures on maps arose, for example to signify

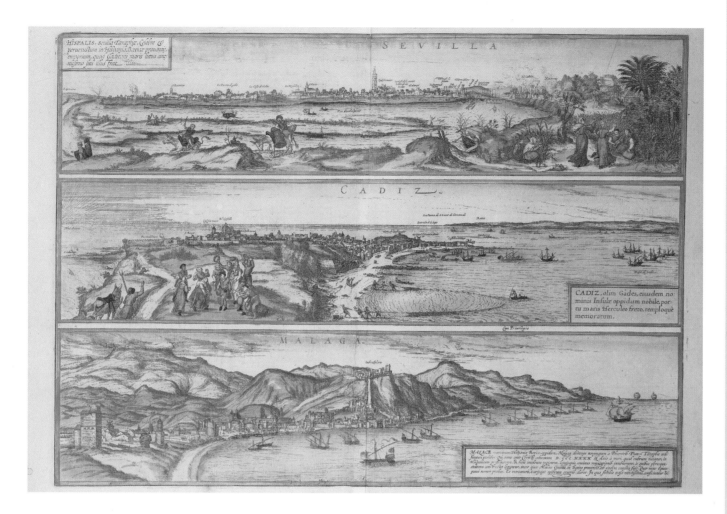

towns. Instead of impressionistic, symbolic or spiritual landscapes, more emphasis was placed on producing a scaled-down image of the physical world.

NEW PERSPECTIVES, NEW PRECISION

Technology was far from the sole driver in the new era of mapmaking. There were also a number of relevant developments in the way in which the world was seen. Far from mapmaking setting the tone, it was painting that was significant, and notably in Renaissance Italy, and, subsequently, in the Low Countries. The linear perspective, then becoming important in landscape and other painting, mirrored cartography in its attempt to stabilise and reify perception and offer a scientific view. In both landscape and maps, there was an emphasis on accurate, eyewitness observation, faithfully reproduced. This process was also seen in urban views, both of buildings and of townscapes.

A key figure was Filippo Brunelleschi who not only provided perspective views of features within Florence in about 1420, but was also responsible for that city's most distinctive skyline feature: the cathedral dome, erected in 1420–1436. Such symbols of civic pride emphasised a town or city's uniqueness, and were to the fore in the so-called chain map of Florence created in 1470.

The use of mathematics to order spatial relationships was one quality that was shared by mapmakers and painters, while cartographic elements dominated many portrayals of landscape. In the mid-15th century, Leon Battista Alberti, an artist and mathematician, set out how to produce a geometric land survey. Other prominent painters also involved in providing plans of cities included Leonardo da Vinci and Raphael. The values and interests of *quattrocento* culture also help to explain the inclusion of new city maps in manuscripts of Ptolemy's *Geography*.

SEVILLE, CÁDIZ AND MALAGA FROM VOLUME I OF *CIVITATES ORBIS TERRARUM* Joris Hoefnagel provided the originals upon which these impressions of Seville (top), Cadiz (centre) and Malaga (bottom) were created. This print is based upon Braun and Hogenberg. Spain's principal port for world trade was Seville. The city exclusively received the so-called *flotas* which carried lucrative treasure and trade goods from the New World and the Indies. Located along a sweep of the Guadalquivir River, Seville's dominant building is shown as the cathedral, completed in the early 1500s, with its bell tower, a former minaret, known as La Giralda. The viewpoint is now the city's barrio de Triana. The Castillo San Jorge, on fire to the right, was then the centre of the Inquisition.

Increasing difficulties with navigation along the river eventually meant that trade moved from Seville to Cádiz. Coastal Cádiz is shown from an oblique projection, or cavalier perspective, which reveals a fortified town with a large church. By the 1600s Cádiz had changed from a fishing town to the gateway to the Americas. Columbus's second and fourth expeditions began in its harbour.

Málaga was the site of a royal arsenal, from where the king of Spain could ensure the safety of his possessions in North Africa. Links with the region lay in the 700-year long Moorish occupation of the city and it is the two-part Moorish fortress complex that dominates the view of Málaga harbour. LEFT

GENOA IN 1481, BY CRISTOFORO DE GRASSI

Painter-cartographer de Grassi copied this in 1597 from a now lost drawing celebrating the departure of the fleet in response to a call by Pope Sixtus IV to liberate Otranto from the Ottomans. The perspective enables the city to be shown enclosed within a circle of hills – and the viewpoint became the classic image of Genoa, seen from the sea. This independent maritime republic relied on its control of seaborne trade, to which the Ottoman expansion (especially the seizure of Constantinople), together with the rise of Venice, had dealt a blow that affected the city's fortunes, in every sense. Genoa is depicted as a tightly packed, almost nondescript, collection of buildings – straining to be near to the activities of the dockside area. The two moles are clearly visible, as well as (to the west) the city's prominent lighthouse. To the centre of the harbour (and the picture) is the Arsenal. Looking eastwards from the Arsenal can be seen the landing stages, beyond which – flying flags – is the Palazzo Ducale. near to the black-and-white striped tower of the cathedral of San Lorenzo. The city's land defences consist of a system of forts sited in the Righi hills, which seem to enfold Genoa in a protective embrace. This location also gave a lookout to sea, for early warning of pirate raids from places such as Algiers. **RIGHT**

In place of the idealised and formulaic representations of cities of the medieval period, came a desire for topographic specificity, so that individual cities looked different. This difference owed much to their relationship with their site, but distinctive internal features were also presented. Moreover, these differences were part of the authority of the image. At the same time, plans were less prominent than images that presented their contents as a visual picture designed to engage and satisfy the eye. This helps explain why the bird's-eye view as a pictorial genre was a major development, one that can be dated to

1500 when an impressive woodcut map of Venice by
Jacopo de' Barbari appeared. In place of an idealised
view presenting the city symbolically as a unit, this
view seeks to provide accurate shape and landmarks in
order to provide an enthralling view of an urban
system, a location of busy life that had a symbolic
value of its own. The city appeared as complete,
readily separated from its surroundings, but separate
not due to walls in an elevated view, but rather thanks
to the distinctive urban activity expressed in buildings.
This approach provided a visual account that was far
fuller than that offered by linear perspective.

SANTO DOMINGO BY BAPTISTA BOAZIO, 1589

This map appeared in an account showing the raid made on New Year's Day 1586 by the English fleet, of seven large ships and 22 others, under Drake against Santo Domingo, the Spanish capital of Hispaniola, and the oldest European city in the New World, founded by Colombus's brother, Bartholomew, in 1496 on the bank of the Ozama River. The Italian cartographer lived in London and may have produced this map from drawings supplied by Drake himself. The plantations, from which the Spaniards derived wealth, can be seen beyond the fortified town. It is through these plantations that the English forces are attacking, overland, before capturing the city and extracting a 25,000-ducat ransom. The grid layout of the colonial city is quite apparent, with the cathedral of Santa Maria le Menor at its heart, the oldest cathedral in the Americas, dating to 1512. **RIGHT**

Greater precision in representation was a theme that was to be linked to scientific developments, especially in optics. City views reflected the application of the new Humanist learning, the glorification of the city as harbinger of the new culture, and the revival of the ancient city republics, notably in the new form of Venice. The city as a whole, a unit that could be grasped visually through pictures and could be adapted by new architectural projects, was a concern of Renaissance rulers, architects and artists. The knowledge of perspective renewed interest in vistas and related harmonies. In Urbino, a major centre of Renaissance painting, the contemporary interest in the creation of perfect, proportionate and whole entities manifested in a fascination with idealised cities, perhaps best known in the form of a perspective view of an imaginary townscape, usually attributed to Piero della Francesca. A perspective view from the ground was different from a bird's-eye view from above, but they shared a common theme of mathematical precision.

PICTURES AND FORMS

Developments during the mapping of countries naturally affected the approach to cities. Whereas earlier mapmakers had been primarily concerned with noting the existence of features, a purpose often achieved by pictures, now there came a stress on recording their accurate shape.

The mapping of major cities posed a particular challenge and opportunity because of the problems of depicting buildings in elevation. Such a presentation combined potent pictorial quality with an architectural essence, but as cities grew there were related difficulties of scale and density. For example, a depiction of London, such as that by Matthew Paris, became less helpful as the city grew in size, which it was greatly to do in the 16th century (including the crowded and thriving southern suburb of Southwark, a snapshot of which is preserved in a hand-drawn map from the 1540s). The woodcut image of London – with rooftops, church spires and city wall – from the 1491 edition of the *Chronicle of England*, originally published by William Caxton in 1480, showed that print could reproduce conventional forms.

Over time, views were to become much more sophisticated, as in Visscher's 'facsimile' of London in 1616, but in the meantime picture elements remained significant in 16th-century maps of the city. Thus the earliest proper map (as opposed to panorama) of London known, attributed to Ralph Agas, and probably surveyed between 1570 and 1605, provides

a series of pictorial elements, including swans and boats on the River Thames and individual people, trees, and deer on land, in the same manner as Richard Lyne's remarkable map of Cambridge, which also has precisely rendered streets and courtyards. In the first known printed map of London, the 'Copperplate Map' of 1559, and the map by Pieter Van den Keene of 1593, there are also individual trees and people. Neither of these maps forsakes precision.

HAND-DRAWN MAP OF SOUTHWARK, C.1542
Southwark is the area on the southern bank of the River Thames. The crossing point into the City of London provided by old London Bridge was the only entrance for all travellers and commecial traffic from the south. This helped to make Southwark an important community in its own right, but most notably because its great southern thoroughfare became the location for many inns and, radiating from there, the adult entertainments centred around the Globe Theatre nearby. Orientated with north to the right, the map shows the many hostelries along Borough High Street, including the George and the Tabard inns (renowned by Chaucer as a gathering point for the Canterbury pilgrims), along with churches, the manor house and other landmarks such as the pillory, stocks and bullring. **LEFT**

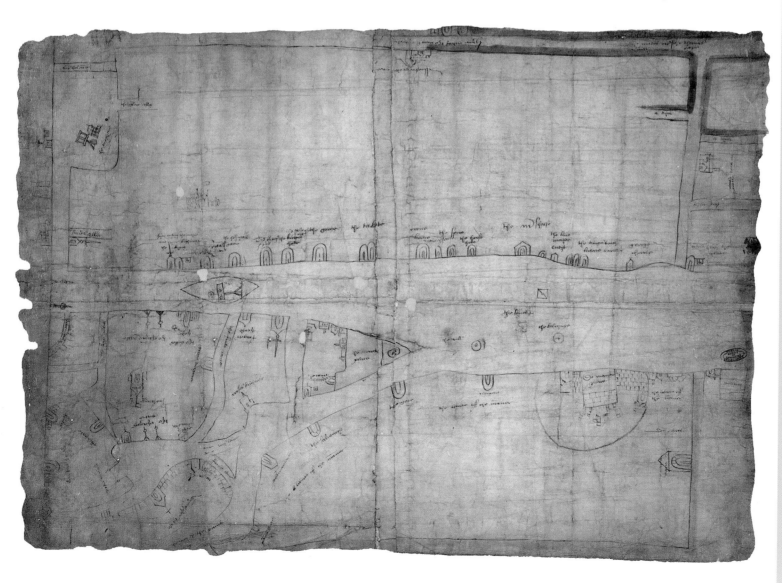

AERIAL VIEW OF VENICE, 1500, BY JACOPO DE' BARBARI (C.1440–1515) This view of the city as a totality is one of the greatest pieces of Renaissance printmaking. It greatly advanced the bird's-eye view as a pictorial genre, not least because De' Barbari's approach was highly detailed and accurate. Canals can be seen which no longer exist (the map is a resource for historians), and the bell tower in St Mark's Square has a temporary roof, following a fire in 1489. De' Barbari used a careful survey of the streets and buildings, plus elevated viewings from different bell towers (using each tower as a survey point, the cityscape was spatially divided into 60 parishes), which means that the map is the product of an innovative proto-digital approach: he created a composite view from multiple partial views. Although highly scientific in its approach, the map is loaded with symbolism. The presence of Neptune and Mercury suggest that the city has been divinely blessed with control of the seas and wealth to be derived from commerce. Also, it has been argued that the distortions created on the left-hand side were deliberately contrived to provide the shape of a dolphin, which had both Classical and Christian symbolism. **RIGHT**

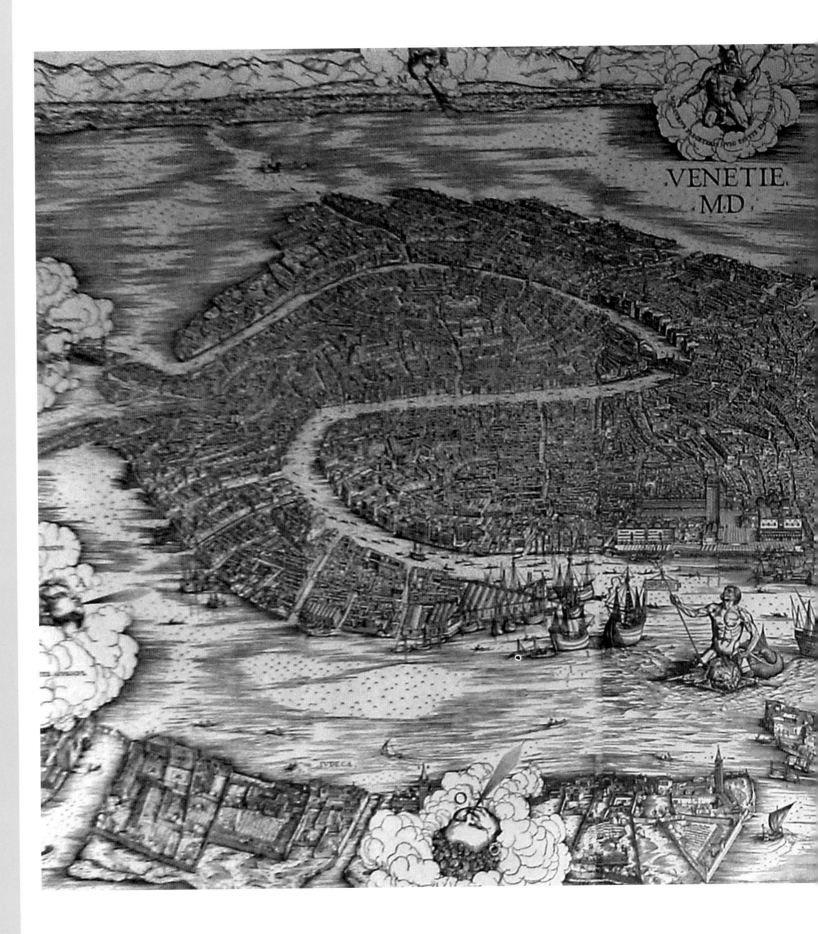

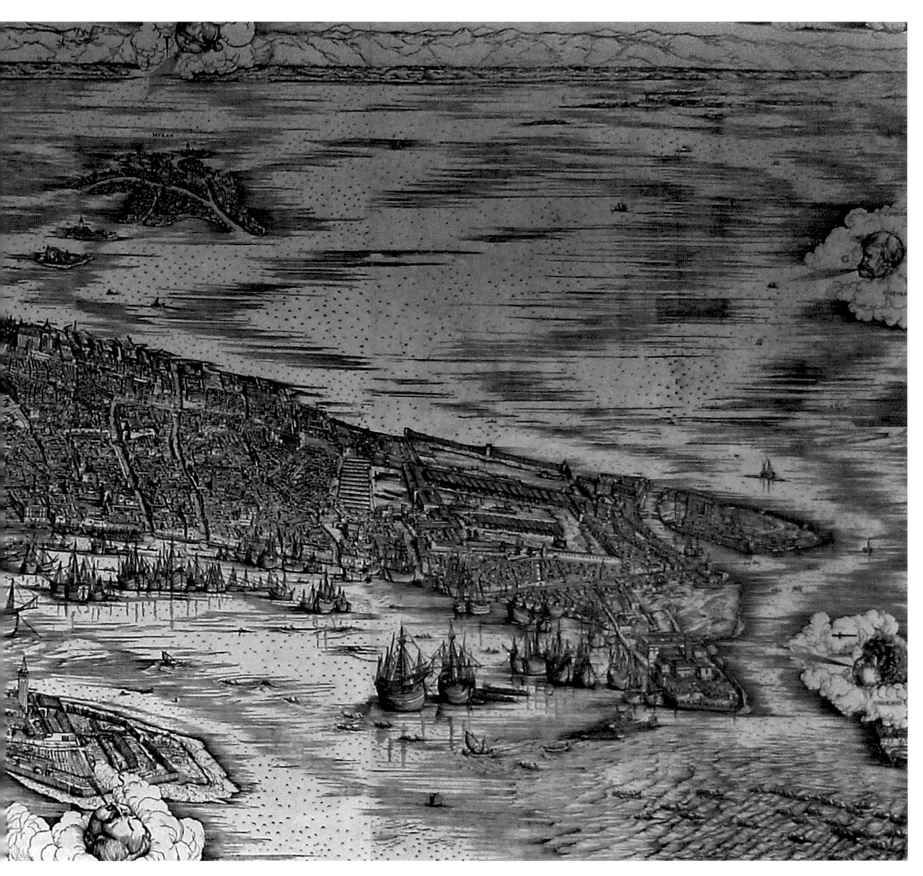

CAMBRIDGE STREET MAP 1574 This is the earliest known street map of the English university city of Cambridge and it was produced by Richard Lyne (his coat of arms are at bottom left), who published it just a year after Christopher Saxton had begun work on his groundbreaking atlas of England. Lyne's map is the earliest English depiction of a town known to have been engraved on copper. With the exception of some later or rebuilt college buildings in Classical style, this detailed map contains much that is recognisable to this day – not least the magnificent King's Colllege Chapel, which was founded by Henry VI in 1441 when this marsh town was still a port. To clear space for his college, the king purchased and levelled houses, shops, lanes and wharves in the centre of the town; he even had a church demolished between the river and the high street. The purchasing and clearances took three years. The main structure of the chapel itself had been completed by about 1515.

In addition to the depictions of the town itself, with its 'yardes', 'greenes' and 'cawseys', Lyn includes some fascinating extraneous details, such as the fisherman near King's College, complete with pointy shoes, and a pig rooting around in Swinecrofte (bottom right). **RIGHT**

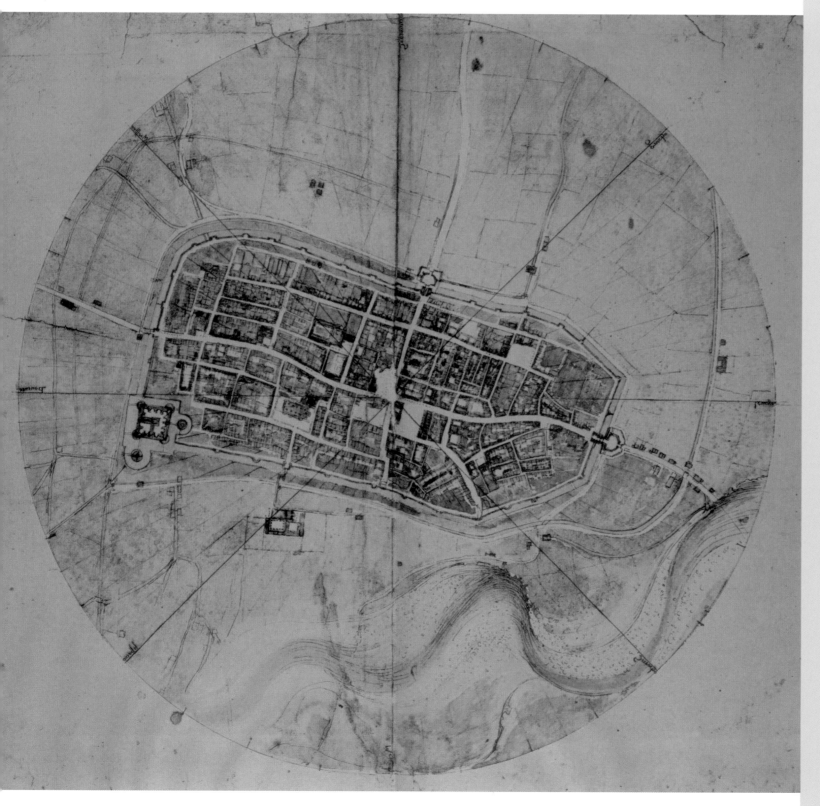

DRAWING OF THE CITY OF IMOLA BY LEONARDO DA VINCI, 1502 Better known to most people through his work as an artist, Leonardo da Vinci (1452–1519) was also a gifted physical geographer, adept at land surveying and cartography. This zenith map was produced for Cesare Borgia, who had asked him to examine the papal fortresses of the region. Leonardo walked the streets and fields to record his measurements and in a highly original presentation, he inscribed the drawing within a circle, the perimeter of which he divided into eight points of the compass, named for the winds, and each, in turn, divided into eight. He famously remarked, 'Let no man who is not a Mathematician read the elements of my work.'

Inscriptions provide the orientations of sightlines to neighbouring towns. Through a heavenly overview such as this map afforded, Borgia gained a useful military tool with which to assess his stronghold and plan its defences. The subtle colour coding blue, green, red identify the nearby Santerno River and the defensive moat, the surrounding fields and the red-tiled houses of the town. **LEFT**

Metropolis

THE CARTA DELLA CATENA SHOWING A PANORAMA OF FLORENCE 1490 This view of Florence is a later pictorial derivation of Francesco Lorenzo Rosselli's (1449–c.1513) famous Pianta della Catena, or 'chain map', woodcut from around 1470–1480, which is a key work in the iconography of this much-loved city. (The name derives from a chain clasped with a lock that framed the top left of the original.) Rosselli's is the first known exemple in the history of cartography that is intended as a complete representation of the dense urban centre with all its buildings and the network of streets and squares. Of the structures, the symbols of civic and religious power stand out: the Palazzo della Signoria, the Palazzo del Podestà and the newly domed cathedral. Near the walls the houses cluster around the churches such as Santa Maria Novella, Santa Maria del Carmine and Santo Spirito. The Arno River, with its four medieval bridges, is lively with human activity animating its waters and banks. Fishing is of particular significance.

The artist is shown in the foreground on the right as he draws the walls of the city onto the sheet, which indicates a starting point at the southwest of the city for this bird's-eye view, delineated in line with the mediaeval approach but with a modern attempt at perspective realism. **RIGHT**

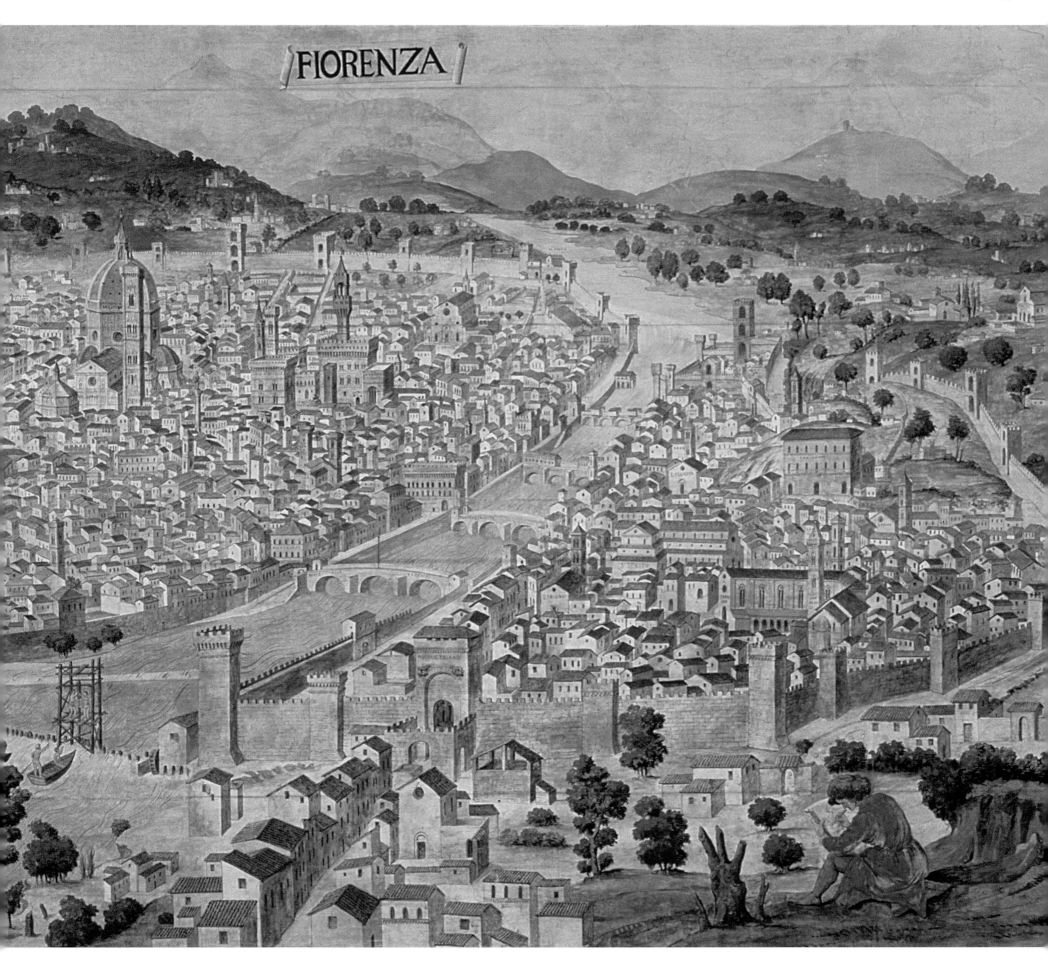

FIORENZA

Metropolis

A BIRD'S-EYE VIEW OF THE GERMAN CITY OF AUGSBURG, 1521 Dedicated to Charles V, who had become Holy Roman Emperor in 1519, and part-funded by Augsburg's wealthy merchant family, the Fuggers (major contributors to Charles's election coffers), this view of the city is from the west and based on scientific surveying, enabling the streets and buildings to be shown in detail. The prosperous city was a key centre of trade and finance, and a major centre of printing, owing to its position along the main route between the eastern Alps in Germany and the Po River valley in northern Italy via the Alpine Brenner Pass, which has served as one of the main entry points into Italy from the north since Roman times. The map, which shows the influence of de' Barbari's plan of Venice, was produced by Jörg Seld (c.1454–c.1527), an Augsburg goldsmith who had other artistic talents and is known to have surveyed the city less than a decade earlier. This is the earliest-known plan view from north of the Alps. **RIGHT**

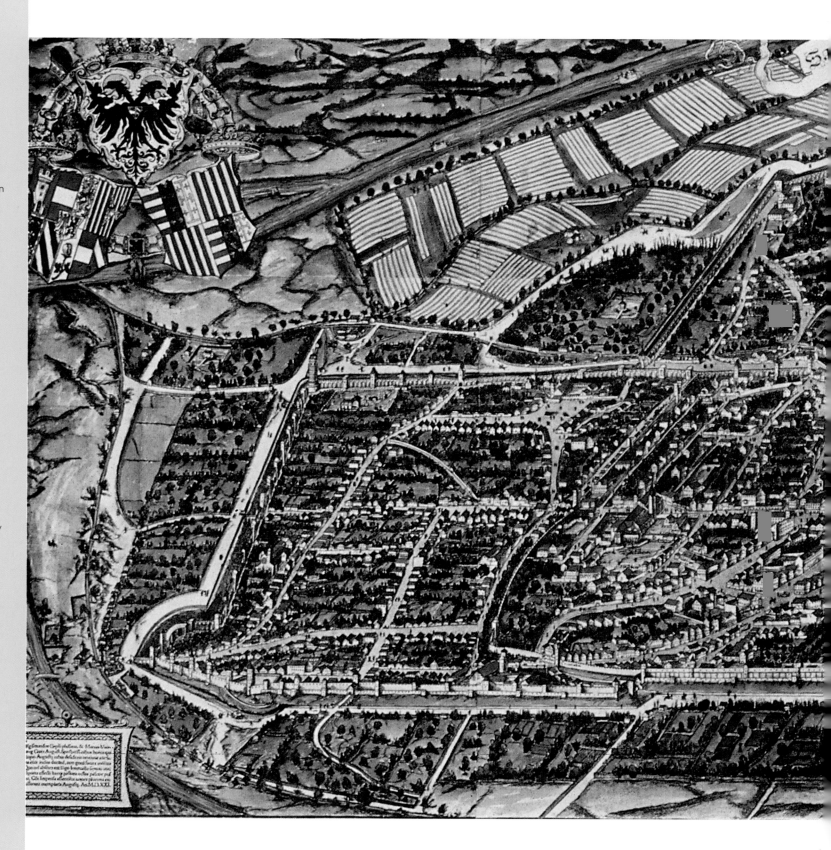

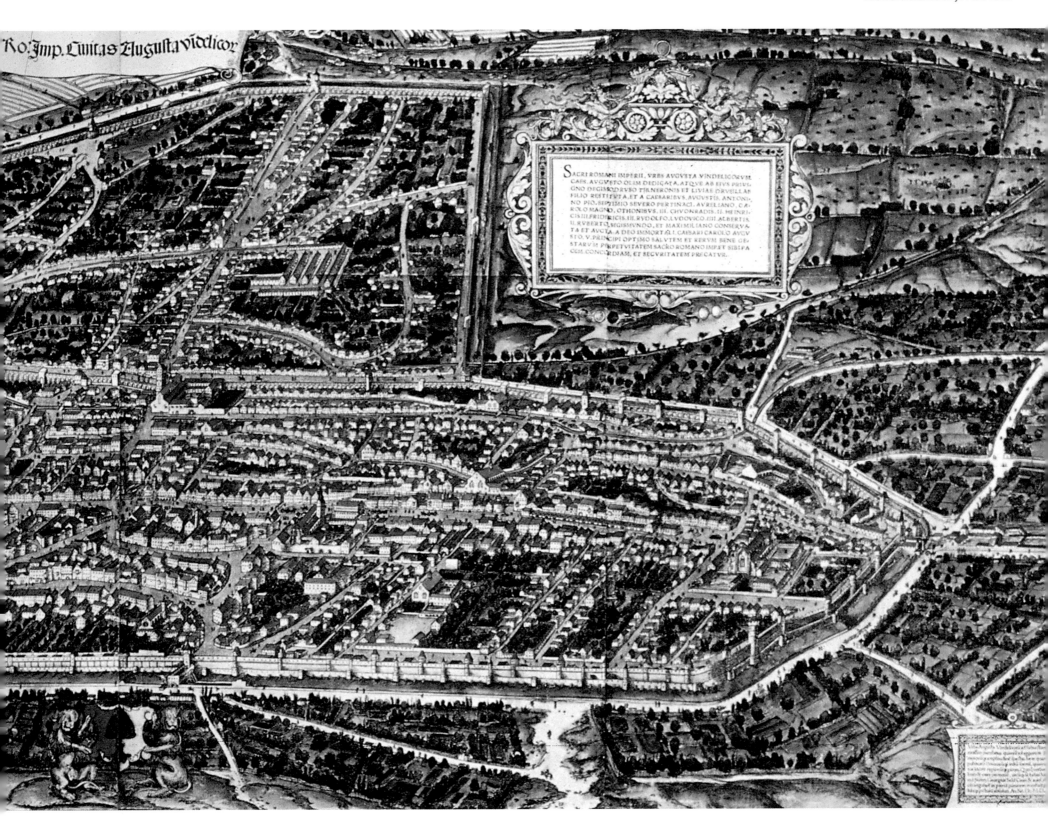

Ro: Imp. Cuitas Augusta vindelicor

SACRI ROMANI IMPERII, VRBS AVGVSTA VINDELICORVM
CAES. AVGVSTO OLIM DEDICATA, ATQVE AB EIVS PRIVI
GNO DECIMODRVSO TIB.NERONIS ET LIVIAE DRVILLAE
FILIO RESTITVTA, ET A CAESARIBVS, AVGVSTIS, ANTONI
NO PIO,SEPTIMO SEVERO PERTINACI, AVRELIANO, CA
ROLO MAGNO, OTHONIBVS. III. CHVONRADIS. II. HEINRI
CIS.III.FRIDERICIS.III. RVDOLFO, LVDOVICO.IIII. ALBERTIS
II. RVBERTO, SIGISMVNDO, ET MAXIMILIANO CONSERVA
TA ET AVGTA. A DEO IMMORTALI. CAESARI CAROLO AVGV
STO. V. PRINCIPI OPTIMO SALVTEM ET RERVM BENE GE
STARVM PERPETVITATEM SACRO ROMANO IMP.ET SIBI PA
CEM, CONCORDIAM, ET SECVRITATEM PRECATVR.

VENICE: Jewel in the lagoon

The great maritime trading centre of Venice was unique among European cities in being preserved in form by its lagoon location. For much of Venice's history, the city had a close relationship with the papacy and trading privileges in cities in the eastern Mediterranean and the East. This ability to link the kingdoms of Carolingian Europe and the Byzantine Empire made Venice and its merchants both affluent and influential.

FROM *THE LIBER CHRONICARUM* BY HARTMANN SCHEDEL, 1493 Nuremberg doctor and book-lover Schedel was the coordinator for a group of scholars responsible for this letterpress–woodcut combination which overcame technical dificulties which had previously hampered layout and printing. The book featured 32 town views like this one, plus 84 smaller views. **LEFT**

FROM *CIVITATES ORBIS TERRARUM* BY BRAUN AND HOGENBERG, 1574 Complete with a key identifying more than 150 locations, and a vignette featuring part of the Doge's procession, this map is accompanied by an interesting commentary. Braun describes a city divided into six districts, and mentions that near the shore stand two huge columns, identifying the place where criminals are punished. He says that the streets of the city are intersected by canals, spanned by wooden and stone bridges, 'of which some 400 can be counted'.

Describing a sea of buildings supported on millions of wooden piles and 180 canals, Braun says that the Rialto is the only bridge over the Grand Canal. Originally made out of wood, in 1591 it was replaced with a bridge of stone. Murano is at the top, to where all Venice's glassmaking industry was relocated in the 13th century in order to prevent fires. The harbour is filled with commercial ships and boats, and local gondolas. **RIGHT**

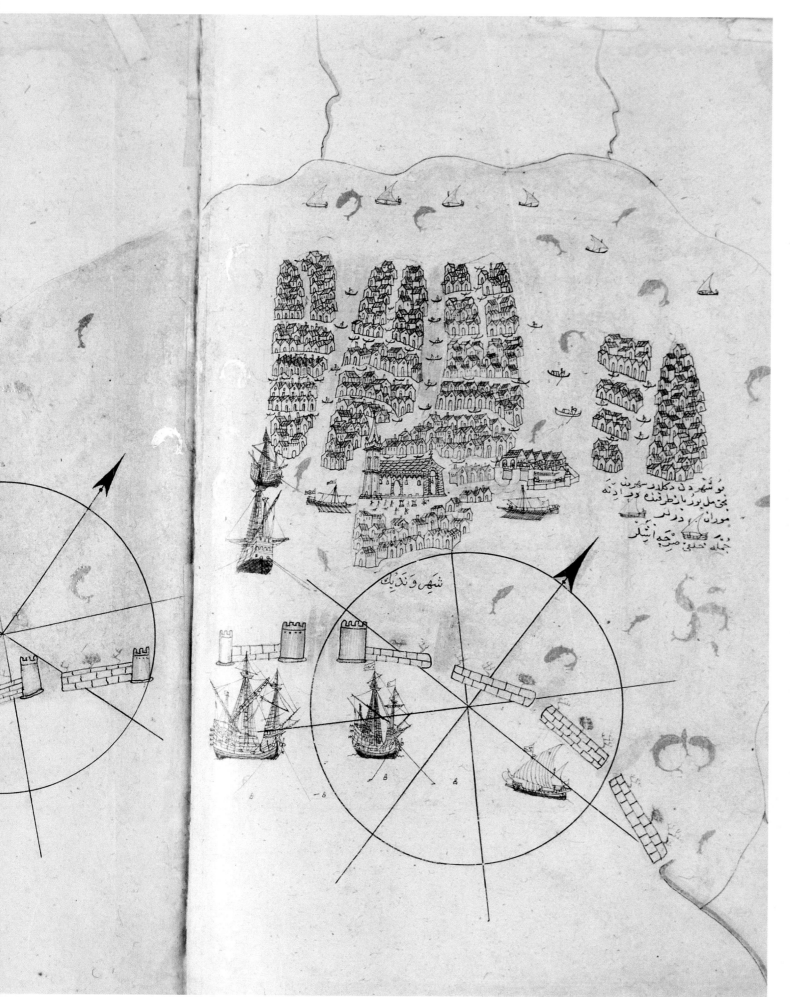

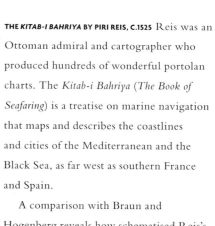

THE *KITAB-I BAHRIYA* BY PIRI REIS, C.1525 Reis was an Ottoman admiral and cartographer who produced hundreds of wonderful portolan charts. The *Kitab-i Bahriya* (*The Book of Seafaring*) is a treatise on marine navigation that maps and describes the coastlines and cities of the Mediterranean and the Black Sea, as far west as southern France and Spain.

A comparison with Braun and Hogenberg reveals how schematised Reis's topographical depiction of Venice is, with lagoon-set islands packed with buildings. Making use of landmarks, such as dominant churches, it gives a realistic sense rather than a scientificallly surveyed rendering.

Although territorial wars did erupt between them, the Venetians and Ottomans were generally able to maintain a mutually beneficial economic partnership in the region, a two-way relationship that involved ideas and valuable goods – and may even have included maps and map templates.

Reis created two versions of his book, in 1521 and 1525, but a third and more richly decorated version was produced in the latter part of the 17th century. The differences seem to be an increasing luxuriousness, with enhanced materials and colour. **LEFT**

CAIRVS, QVAE OLIM BABYLON, AEGYPTI MAXIMA VRBS.

CAIRO BY SIENESE MAPMAKER MATTEO FLORIMI, C.1600 Based on Braun and Hogenberg's depiction of the city, Florimi's perspective looks down onto the city from higher ground across the Nile. The engraving appeared in *Quae Olim Babylon Aegypt Maxima Urbs* and it shows the famous pyramids, Sphinx and an obelisk in rather stylised form to the right of the print. Braun and Hogenberg's image was, in turn, probably based on a woodcut (c.1549) by Matheo Pagano whose information seems to date to earlier because of the position of the city's aqueducts. **ABOVE**

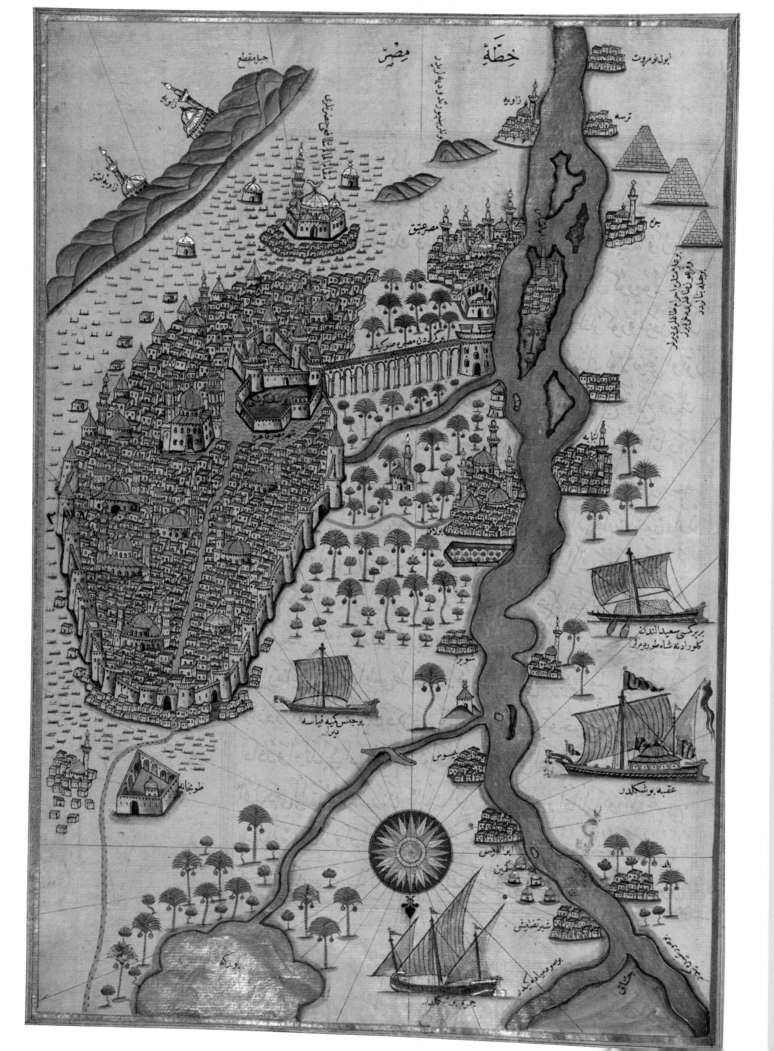

CAIRO, FROM PIRI REIS'S *KITAB-I BAHRIYA*, C.1525
This beautifully coloured map results from Piri Reis voyaging inland, up the Nile to Cairo. Giza is on the West Bank of the Nile, which tells us that south is at the top and north is at the bottom. The Muqattam hills to the east of Cairo are depicted at top left. The main point of interest is the citadel of Cairo, begun by Saladin in 1176. The citadel was enlarged by the Mamluks into a complex with two enclosures and it was further expanded by the Ottomans after they conquered Egypt in 1517. Rebellions encouraged the reliance on the citadel. Most of what can be seen in today's citadel dates from the Ottoman period. From the Nile, an aqueduct leads to the citadel, with its two enclosures. In the bottom right corner of the latter is a palace. Visible to the left of the bottom enclosure is the domed mosque–madrassa–mausoleum complex of Sultan Hassan. Some of the verdant area around the aqueduct corresponds roughly with the modern residential district of Garden City. LEFT

Metropolis

MAP OF ALGIERS FROM *CIVITATES ORBIS*
***TERRARUM* BY BRAUN AND HOGENBERG, 1575**

Algiers and Tunis were major bases out of
which galley fleets raided widely in the
Mediterranean and beyond to contest
control of the region with the Christian
powers. These Barbary Corsairs were
privateers, encouraged to conduct raids
against Christian territories, in search of
plunder and slaves, which were
legitimised by *jihad*, while the state took
some of the benefit. In the 1540s there
were attacks on Catalonia and the coast
of Italy. There were even raids in British
waters and several major expeditions
were launched against Algiers, which
were either defeated by poor weather
or Algier's defences. The map shows the
fortress-like layout of the harbour and
the proximity of the town, with five
mosques visible, including the Grand
Mosque built in the 11th century. The
mole extends from the city gates out
into the harbour. Algiers fell to the
French in 1830. RIGHT

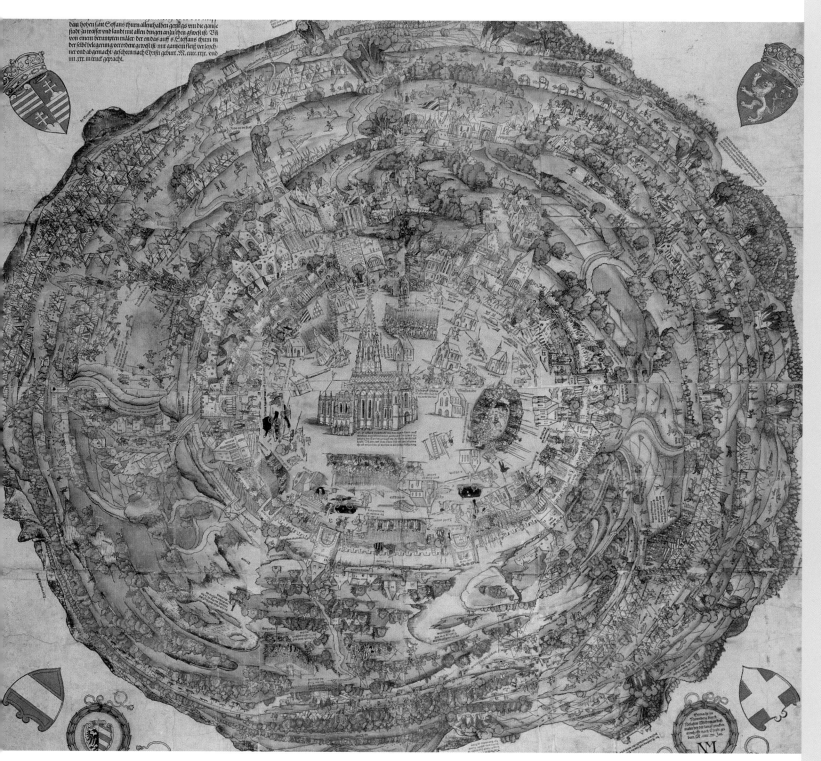

CIRCULAR MAP OF VIENNA BY HANS SEBALD BEHAM, 1530 Published by Niklas Meldemann in Nuremberg shortly after the Ottoman siege of 1529, this woodcut print offers an informative topography of old Vienna. The map was sketched from the spire of St Stephan's cathedral, the circular design emanating from that central point, which was the headquarters of the defending commander, German mercenary Niklas von Salm. In the surrounding countryside are the besieging forces of Süleyman the Magnificent. The city's 300-year-old walls are flattened out in the view so that the inner surfaces are visible, thus enabling defensive armaments to be depicted. Salm conducted a vigorous defence, blocking the city's gates, reinforcing the walls with earthen bastions and an inner rampart, and levelling any buildings where it was felt to be necessary for the good of Vienna.

Having withstood the assault, Vienna afterwards built massive, purpose-built encircling fortifications – which were eventually put to the test when the Ottomans returned in 1683. **LEFT**

CONSTANTINOPLE:
Where West encounters East

By far the greatest city in Christendom, the size, beauty, wealth and magnificence of Constantinople were beyond the conception of most western Europeans in the Middle Ages. Various factors account for the city's decline, but the sacking during the Fourth Crusade (1204) fatally weakened it. While the Venetians enriched themselves with loot from the city, their actions hastened the end for the capital of the Christian East.

FROM *BEYAN MENAZIL* BY MATRAKI NASUH, 1537 This view of Constantinople is in debt to the earlier map by Buondelmonti (see page 24). But having begun on that basis, the viewpoint of the main peninsula has been tilted anti-clockwise by 90 degrees, resulting in Constantinople being to the right and Galata to the left. Although not overtly realistic, Nasuh seems to have used a form of colour coding for building materials: roofs are grey for lead and red for tiled; structures are white–grey for stone masonry and yellow for stuccoed-brick. Types of building can be identified too: mosques, tombs and hamams. **RIGHT**

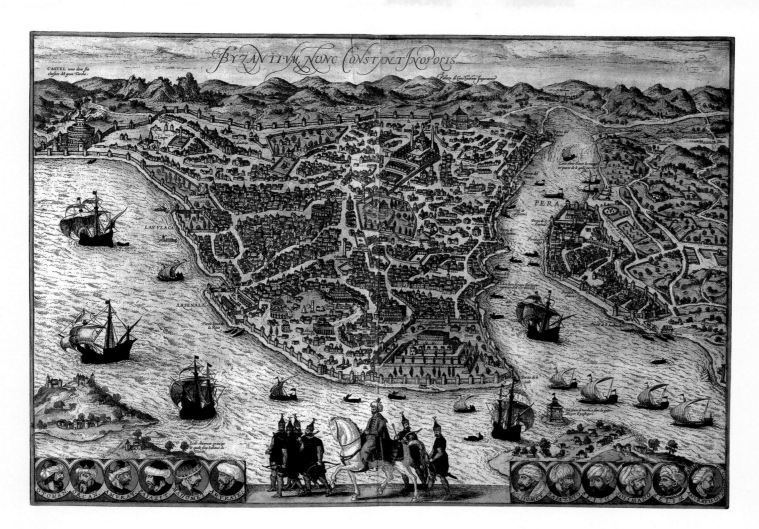

FROM *CIVITATES ORBIS TERRARUM* BY BRAUN AND HOGENBERG, 1576 This bird's-eye perspective from the east indicates Byzantium's process of transformation, with its prominent Ottoman sultans. The orientation also emphasises (in the foreground) the Topkapi Palace on Seraglio Point, built in c.1459–1465 on the site of the old Byzantine acropolis. Hagia Sophia already has minarets. **LEFT**

FROM *THE LIBER CHRONICARUM* BY HARTMANN SCHEDEL, 1493 A view of what had been, since 324, the great imperial centre of Eastern Christendom. Among the city's imagined features is an accurate detail in the formidable double land walls. **BELOW**

THE SEVEN CHURCHES OF ROME, AFTER ANTONIO LAFRERI, 1575 This thematic map exemplifies a cartography focused on a particular topic, wherein the rest of the cityscape slips from view. These symbolic seven are the major basilicas that constitute a well-defined route for pilgrims. In the foreground is the undomed St Peter's. **ABOVE**

MAP OF VATICAN AREA, BY BARTOLOMEO FALETI, 1561 Castel Sant'Angelo fortress incorporates the tomb of the Roman Emperor, Hadrian. An imposing sight, it gives a sense of a city within a city to pilgrims on their way to St Peter's. Detailed topographical views of Sancta Roma (Holy Rome) such as this catered for both the real and virtual tourist demand from the earliest days of printing onwards, reflecting how Rome was seen and mapped – spatialising it for them – through the eyes of the many travellers who went there. ABOVE

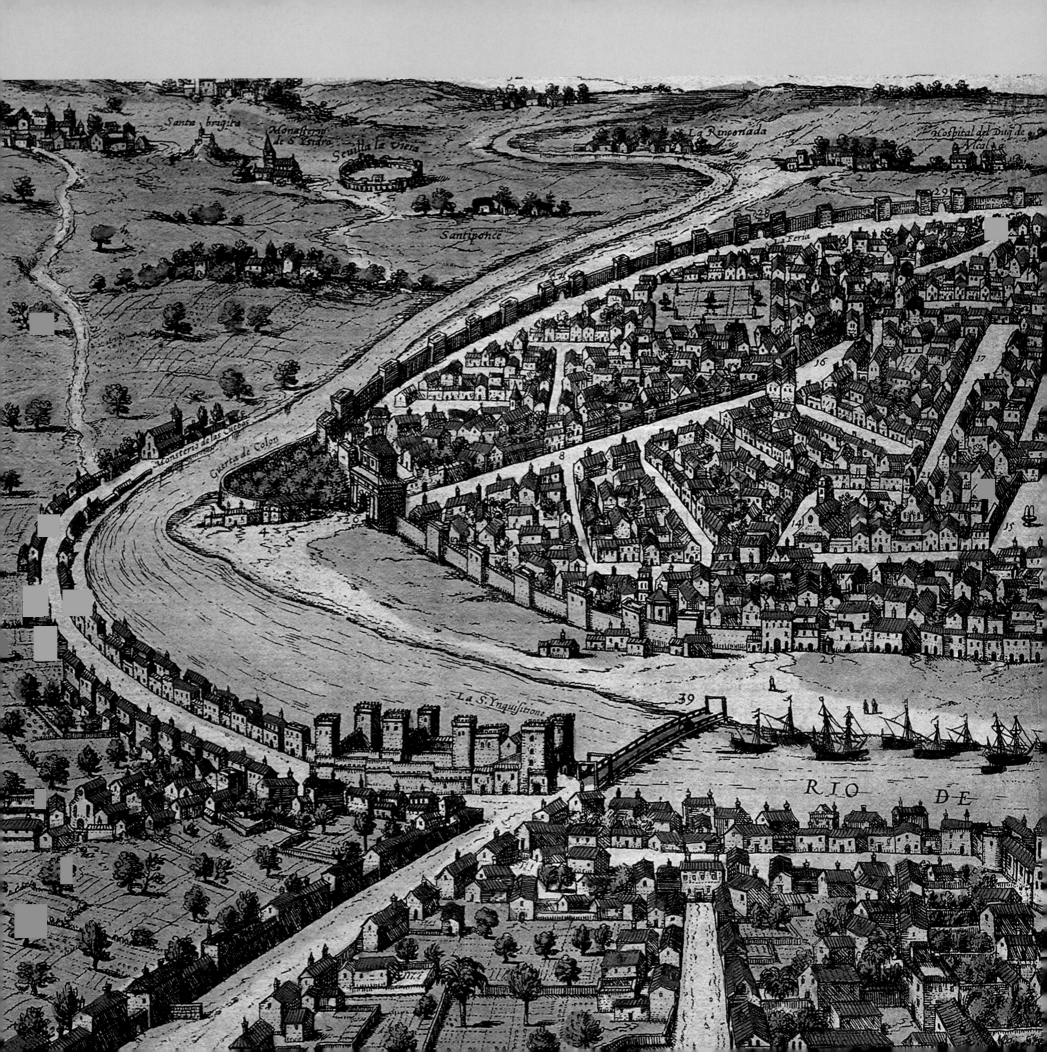

Santa brigita

Monasterio de S. Isidra

Seuilla la Vieia

Santiponce

La Rinconada

Hospital del Duq de Alcal

Monasterio de las Cuebas

Guerta de Colon

La Feria

La S. Ynquisitione

RIO DE

28

29

16

17

8

14

27

15

26

4

23

39

GVADALQVEVIR

China and India remained the world's most populous places in the 17th century and, therefore, the location for such major cities as Beijing and Nanjing in China, and Agra, Delhi, Lahore, Bijapur and Golkonda in India. Elsewhere in Asia there were important trading entrepôts, such as Johor in southern Malaya, that constituted significant cities.

NEW WORLD SETTLEMENTS

However, the further expansion of the European maritime empires ensured that it was cities on the other side of the world – where this globally emerging trade was being organised – that were of increasing world significance in the 17th century. In addition to Lisbon and Seville, both of which remained very important because of their key roles in a global trading nexus, elsewhere in Europe the more northerly cities of Amsterdam and London came to have a greater part to play in world commerce durig this period.

In addition to that, new cities were founded by Europeans overseas, notably in North America and the Caribbean where Quebec (1608), New Amsterdam (1614, later New York) and Port Royal (1655, Jamaica) were all established. Before the selling of a vast North American territory to the USA in 1803 with the Louisiana Purchase, New France had two key geographical points: New Orleans in the south and Quebec in the north. The latter's strategic, clifftop location on the St Lawrence River was intended to give France control of trade and communications with the northern interior. New Orleans was founded in 1718 as a shipping centre that commanded the mouth of the Mississippi–Missouri river system, which drains much of the continental interior.

Further south, in 1609 Englishman Henry Hudson reported to his employer, the Dutch West India Company, that there was a protected anchorage near good farmland on the southern tip of Mannhata island – the harbour where Italian navigator Giovanni de Verrazzano had reported a deepwater anchorage during his 1524 expedition. By 1626 the approximately 22,000 acres (9,000 hectares) acquired from the Lenape people for 60 guilders was beginning life as a Dutch trading post, the settlement of New Amsterdam.

The natural harbour of Port Royal in the Caribbean was a major British trading post in the 17th century – partly thanks to the predations against the Spanish treasure fleets – but a devastating earthquake in 1692 put an end to the city's fortunes and led to the founding of Kingston, Jamaica, across the bay.

In 1652, Cape Town in southern Africa was founded to supply the ships of the Dutch East India Company en route to and from the Indian Ocean. The English East India Company then set up a permanent trading site at Sutanuti, which became the basis for the later city of Calcutta. Already founded cities on either side of the Pacific Ocean, such as Lima and Manila, also continued to increase their populations.

Each of these cities were centres of trade, government and power, and the intermediary points between imperial metropoles and their overseas presence. As a consequence, they were the foci for any overseas conflict and were fortified accordingly. Thus, the Portuguese erected a significant defensive bastion at Malacca to secure their regional interests, which enabled them to defend the city for more than a century, until in 1641 the Dutch wrested control from them.

URBAN DEVELOPMENTS

Any discussion of the importance and growth of cities is made difficult by problems of defining what a city, or indeed a town, was. The legal definitions of the

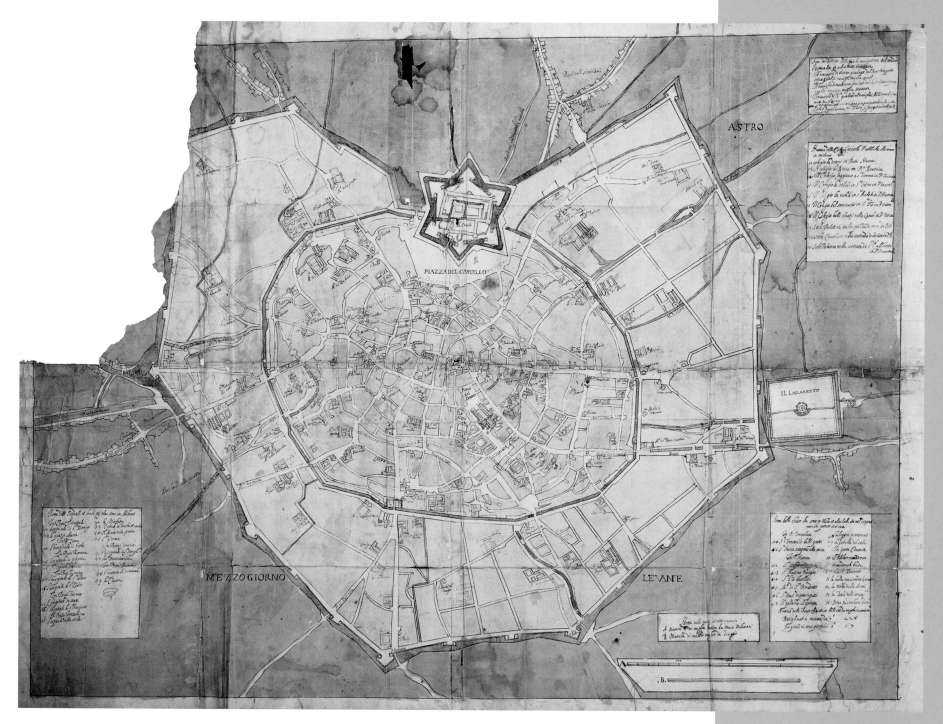

period paid scant attention to size or function. In parts of France, all built-up areas surrounded by walls and provided with royal justice met the criteria, irrespective of their size or function. In Poland, it was a case of legally incorporated institutions, again without regard to size or function. A settlement without an owner was the key criteria in Russia. Although they differed greatly in age, size and other factors, such places were subject to the same regulations, and contrasted legally with settlements that were the property of an individual or an institution.

Such definitions were of limited value functionally. Moreover, there was no agreed contemporary definition of a town, as well as no legal classification valid across an entire continent.

There are also problems with imposing modern notions. To follow one modern definition – that of the city as a creator of effective space, a nexus whose most crucial export was control and the locus of a characteristic urban lifestyle – would be to ignore the role of agrarian activities in many cities in the past.

The use of legal classifications was complicated by the frequent failure to define one urban area as only one city (or town). An urban status was essentially a matter of legal privilege rather than population size or economic function, and there was no particular reason why this status should be rationalised in accordance with contiguous urban areas so that they were treated as single cities.

This had significant implications for mapping. Enclaves of distinct status frequently existed within contiguous urban areas. Urban status was a particular problem in the case of suburban or extra-mural (outside the walls) developments. If cities of the period present problems of definition, their population size is not always easy to assess, particularly as many statistics for the past were calculated in terms of households, not people.

CITIES AND STATES

Italy, although dominated by Spain in the 17th century, contained a number of independent states that were focused on cities. In the northern region of Lombardy the duchy of Milan was an example of despotic patronage in the shape of the Sforza family

MAP OF MALACCA, MALAYSIA, C.1620S The city of Malacca, or Melaka, situated in a key position in the Straits of Malacca, developed as a funnel for trade between Malaya and Sumatra, and thus between East and South Asia. It exemplies the rise of cosmopolitan coastal cities in the region, benefitting from links to India and China. Because such locations were possible choke-points on trade and naval movements, they often became fortress-bases, which was the case with Malacca after its capture by the Portuguese in 1511. Afonso de Alburquerque replaced the local palace with his bastion overlooking the harbour, which symbolised Portuguese control for the next 130 years, until the Dutch captured the port in 1641. The fortaleza de Malaca, or Melaka fort, is shown in the centre foreground. A covered wooden bridge over the Malacca River links the fort to the native city on the bank opposite, the space between them becoming a city centre and market. Malacca thus had two distinct racial sections: the fortified part containing Europeans and the open indigenous part where different trading nationalities mingled. At the top of the picture is an inner fort containing a church. RIGHT

THE SEVENTEEN PROVINCES OF THE NETHERLANDS, 1648, BY JOANNES VAN DOETECUM
Based on an earlier engraving by Claes Jansz Visscher, c.1610, called 'Leo Belgicus' this map of Netherlands unity was a common motif during the Twelve Years' Truce (of the Eighty Years' War), intended to maintain a Netherlands identity in opposition to Spanish domination. Rooted powerfully in civic pride, Dutch identity and aspirations for renewed unity are expressed in the map through the roaring allegorical lion (picked out from the outline of the 17 states of the Netherlands) alongside the great mercantile cities of the northern and southern Netherlands that were for many the country's core of identity – none more so, arguably, than the great port of Amsterdam. **LEFT**

whose passion for military engineering left a permanent mark on the city. Milan's distinctive canal network also received a significant enhancement with the extension of the Naviglio Grande to link the city to Lake Maggiore (where the marble to build the Duomo was sourced).

City-state republics such as Venice and Genoa continued to have a role and urban republics were widely applauded in the 16th and 17th century West as models of good government and civic virtue. They could be seen to resonate with Classical values that were closely associated with republicanism. Public virtue could be presented as a republican characteristic, the product of states with a 'balanced' constitution.

The United Provinces (Dutch Republic) appeared to epitomise this as a form of civic federation.

The greatest Dutch city was the sea-traffic trading centre of Amsterdam. Founded as a fishing village in about 1100 and receiving its first charter in 1275, Amsterdam had become a manufacturing and trading town which entered its Golden Age in the 17th century. It was then that it had its major period of canal building and the vast town hall was completed in 1665 as a display of civic pride and urban independence.

Nevertheless, despite the powerful desire of many cities to run their own affairs as independent entities, territorial states might appear to be more effective providers of protection by ensuring that cities in these states, unless they were in frontier zones, did not require modern (and costly) fortifications or large militias. Many such states therefore developed as

MAP OF BATAVIA, 1682, PUBLISHED BY AMSTERDAM MAPMAKER JACOB VAN MEURS From 1619 onwards the Dutch developed Batavia as the centre of Dutch power in the East Indies. Built like a Dutch planned city, with blocks delineated by canals, squares and tree-lined streets, the city was dominated by an important fort (*kasteel*) on the bank where the original sultanate of Jayakarta had been.

Completed by approximately 1650, the city was a walled and fortified town opposite the fort, with the Ciliwung river feeding a defensive moat and an irrigating system of canals for the surrounding orchards and ricefields. The fort dominates the harbour from the west bank of the river.

Wealthy Dutch merchants built themselves fine houses – helping to earn Batavia the nickname 'Queen of the East'. However, the climate and the stagnant water in the canal system also made Batavia an often disease-ridden stopover for sailors returning to Europe from further east. Outbreaks of tropical diseases such as malaria also contributed to the growth of villa suburbs to the south of the city. **RIGHT**

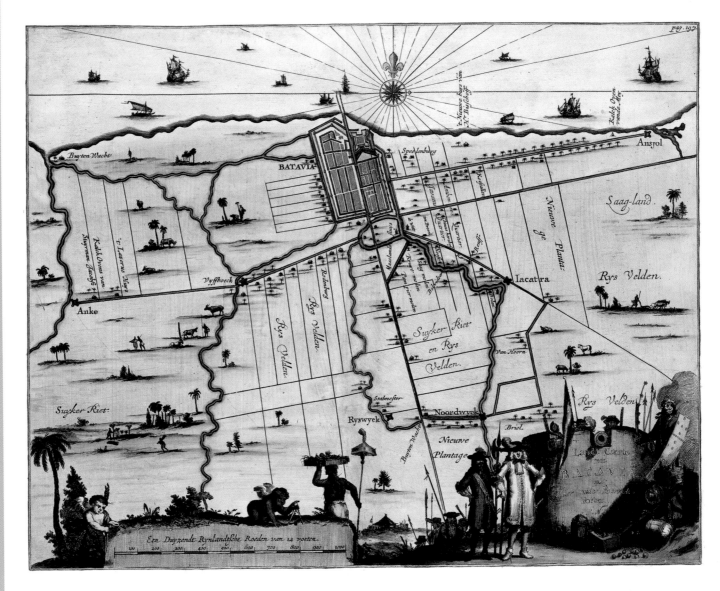

combinations of the landed power of rulers and rural élites married with the financial resources and mercantile interests of urban élites.

This cooperation was not seen elsewhere, largely because of ethnic and social differences between rural and urban élites. Thus, the Ottoman capital was now in the great port city of Constantinople, but the Muslim landed élite had scant affinity with the often Christian or Jewish mercantile élites in trading centres such as Alexandria, Smyrna (Izmir) and Salonica (Thessalonica).

The six volumes of Braun and Hogenberg's *Civitates Orbis Terrarum* were followed in the 17th century, by numerous plans and views, both for Europe and for areas under Western control. Thus, in 1659, a large-scale map of Batavia, the Dutch base in Java was published. In 1603 the Dutch had established a trading

base at Jayakarta (modern-day Jakarta on the Indonesian island of Java), located on the bank of the Ciliwung River. From 1619, the town became the centre of Dutch power in the East Indies after it was stormed and the forces of Banten, the local sultanate, were defeated. Renamed Batavia (after Dutch ancestors in antiquity), the town subsequently survived two sieges by Sultan Agung of Mataram, another Javan power, which signalled that the Dutch could withstand the challenge from powerful local interests.

The growth of this Dutch commercial world helped ensure that Amsterdam became the leading centre for mapmaking. Amsterdam drew on a global system of trade. Moreover, Dutch cartography was linked to Dutch painting, as both attempted the description of physical reality.

PLANNING PARIS AND LONDON

As a rival to Amsterdam, Paris became a significant centre for mapmaking, but much of the patronage was royal and aristocratic, and the culture of cartography was different to that in the Netherlands. The expansion of Paris in the 17th century was marked by the building of royal palaces and major projects, such as the Luxembourg Palace, the Palais-Royal, the Institut de France, the Observatoire (Observatory) and the Invalides. London, which became a more important centre of mapmaking from the late 17th century, was more similar to the Dutch model.

The Great Fire of London destroyed much of the city in 1668, burning 373 of the 448 acres (181 hectares) within the ancient city walls and another 63 acres (25 hectares) to the west. In the aftermath, King Charles II issued a declaration promising that London would be rebuilt better than ever, and would 'appear to the world as purged with the Fire … to a wonderful beauty and comeliness, than consumed by it'.

Adding a dimension that could not be readily captured by plan maps (as opposed to views), he also declared that a handsome vista would be created on the riverbank by banishing smoky trades that might

QUÉBEC BESEIGED Produced in 1705, this map (oriented with north to the left) by an unknown artist depicts the events of several years earlier when a British fleet laid siege to Québec, founded in 1608 as the capital of New France by Samuel de Champlain atop a plateau overlooking the St Lawrence. The map provides a fine view of the fortified city's commanding location and has a valuable numbered key. Protected by the Fort of St Louis, the city's harbour facilitated the extensive fur trade and from the earliest days the topography of the site meant that the city was spatially zoned: an administrative and religious centre developed in the upper city, while a business and maritime centre evolved in the lower city.

Québec is the oldest non-Spanish city in North America, and its preserved defences make it the only walled city north of Mexico. **LEFT**

NEW AMSTERDAM, 1660, BY JACQUES CORTELYOU
Known as the Castello plan (because the original was stored in a Medici castello in Florence), this is the earliest surviving Dutch map of the settlement of Nieuw (New) Amsterdam, executed in enough detail to recognise typically Dutch steep-gabled buildings and many orchards and garden plots. New Amsterdam was not the first Dutch settlement in North America, but the advantages of its location made it immensely valuable. Peter Minuit and his successor governors knew that expanding Dutch commerce was their primary task, and the wealth from the lucrative trade in fur supported the early development of the city. However, as early as the 1630s there were complaints that too much of the city was given over to sating the wants of sailors stopping over there. Although there were clashes with local Native Americans, the settlement gradually moved northward, laid out farms, and expanded trade with the rest of New England and the outside world. Although after losing it in 1164, the Dutch briefly reoccupied the colony in 1673–74, its destiny was now firmly linked to London. **RIGHT**

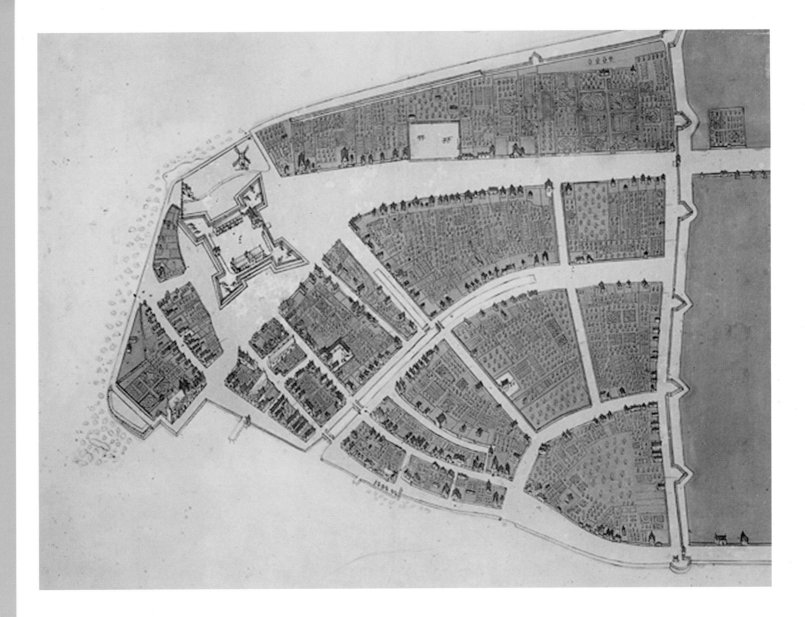

mar the view. John Evelyn, Peter Mills, Christopher Wren, Robert Hooke, Richard Newcourt and Valentine Knight then produced plans for rebuilding London as a consciously modern project and to a more regular plan, which was the pattern followed in other leading European cities, and one that reflected the favour of the period for dramatic, long streets and rectilinear town plans.

Proposing interconnecting squares and piazzas, Evelyn urged the value of zoning, especially the removal of noxious trades from areas of polite habitation. Wren's design showed boulevards replacing the narrow alleys and streets of the medieval city. He proposed two central points, the Royal Exchange and St Paul's. In Wren's scheme, ten roads were to radiate

from the former, while a piazza in front of St Paul's was to be the focus of three key routes in the western part of the city. Wren's new London was laid out on a grid dependent on the major through-routes, with piazzas or rond-points playing a key organisational role. The river was to be faced by a grand terrace. Hooke and Newcourt each proposed a regular grid.

In the event, several factors ensured that the rebuilding was rather more piecemeal and unplanned than had been intended. The drive for a rapid rebuilding was paramount, not least the need to tap the resources that could be readily raised by existing property-owners. As a result, boundaries did not greatly change. The sole new street created a new route from Guildhall to the river.

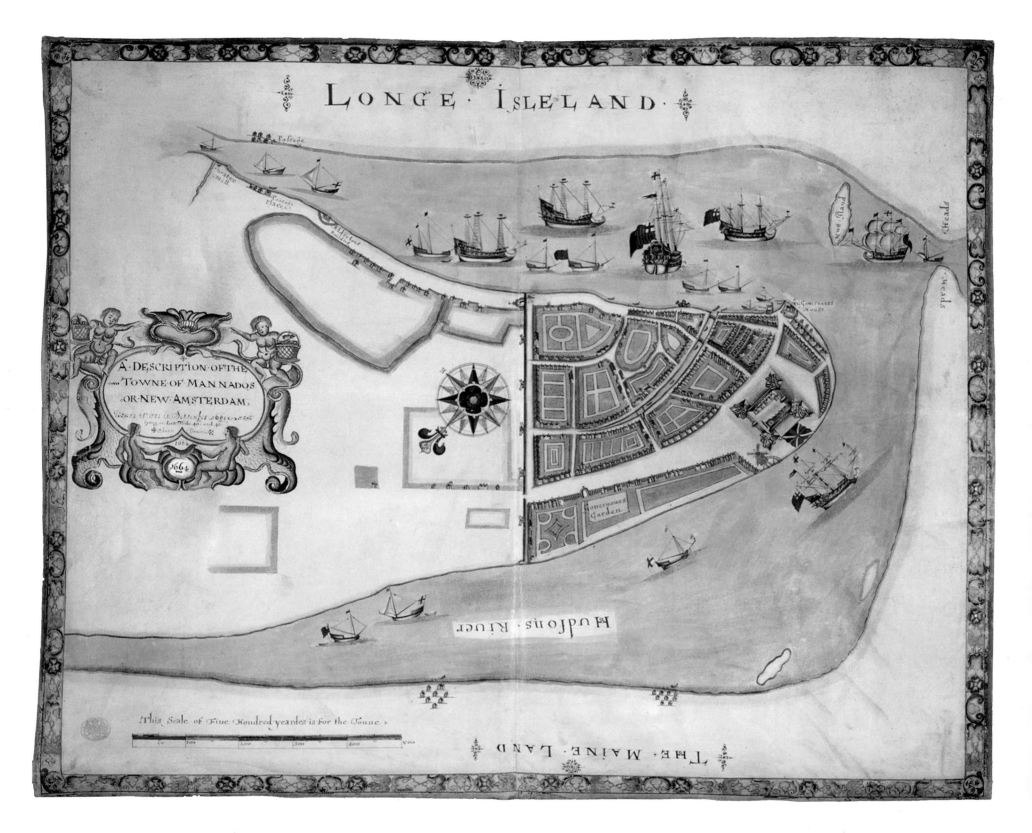

NEW AMSTERDAM, SEPTEMBER 1664 Despite the use of the name 'New Amsterdam', the English ships in the harbour probably date this map to September 1664, after Peter Stuyvesant, director general of New Netherland, surrendered to an English fleet sent by James, Duke of York, and it had been renamed in his honour. Stuyvesant's attempts to impose order on the colony left him with few allies when the fleet arrived. The map resembles the Castello plan, apart from the 180-degree change in orientation, with the East River at the bottom. The fort that gave its name to Battery Park is shown and the windmill nearby still features in New York's city seal. The long straight route that leads northwards from Bowling Green, adjacent to the fort, was known as Heere Straet and is known today as Broadway. The city's northward limit is marked by the palisaded ditch along which Wall Street runs today. William and Broad (with its canal) streets run down to the tip of the island. **ABOVE**

MAP OF THE CITY OF LONDON BY WENCESLAUS HOLLAR, 1666 The light area north of the Thames shows the extent of the area destroyed in the four-day Great Fire. The inset map at the lower left covered Westminster and the suburbs as well, revealing the proportion of the city destroyed: 373 of the 448 acres within the ancient city walls and another 63 acres to the west, destroying more than 13,000 houses, nearly 90 churches, St Paul's cathedral and much of London Bridge. The map is an interesting combination of aerial view (which shows the surviving buildings, such as the Tower) and plan (to depict the empty streets). RIGHT, TOP

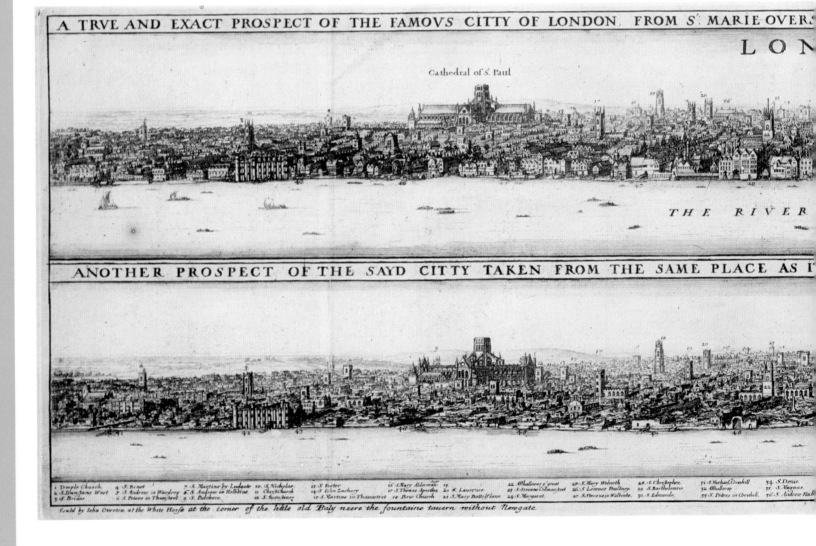

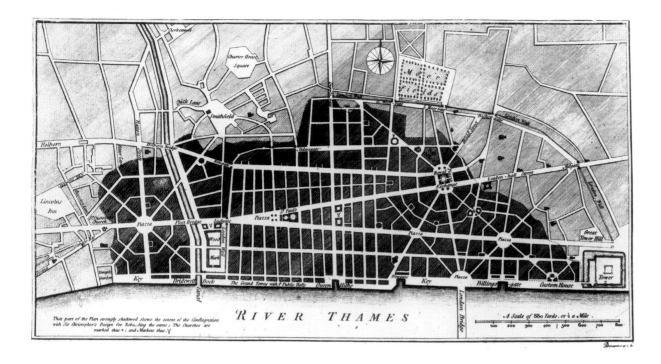

New horizons, new worlds: 1600–1700

CHRISTOPHER WREN'S PROPOSAL FOR LONDON AND HOLLAR'S VIEW BEFORE AND AFTER THE GREAT FIRE OF 1666 The dark-shaded part of Christopher Wren's plan shows the area of devastation created by the catastrophic Great Fire of London. Wren's proposal was largely ignored on financial grounds, though he did oversee the rebuilding of many churches. His idea was to replace the warrens of old, narrow streets with modern boulevards geometrically linked to piazzas. The grand formal street plans (such as Pierre Bullet's for Paris) that Wren had studied on the continent are a clear influence. He also suggested a Thameside quay, from Bridewell to the Tower, with warehouses instead of the ramshackle wooden wharfside buildings. Hollar's contrasting before-and-after panoramic prospect is the view from the steeple of Southwark cathedral, with a key to buildings. **LEFT, TOP AND BOTTOM**

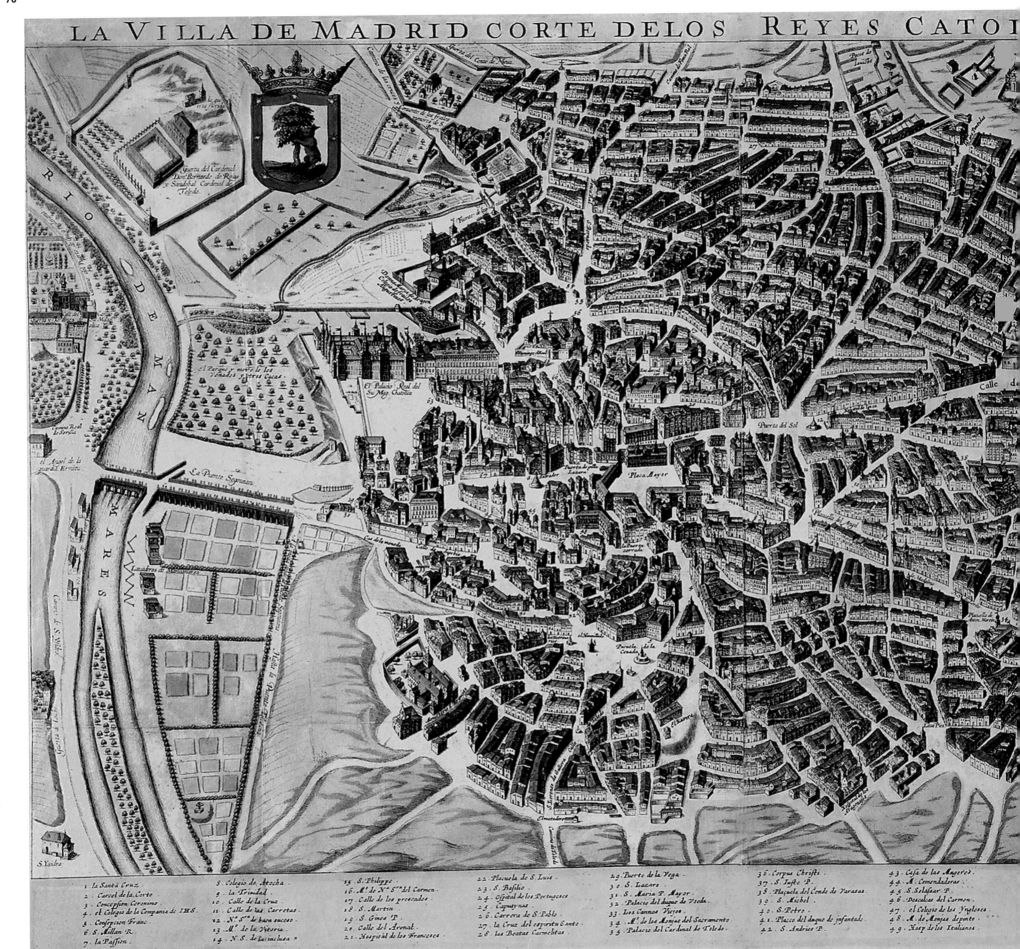

RIO DE MAN

MARES

S. Yndro

1. La Santa Cruz.	8. Colegio de Atocha.	15. S. Philippe.	22. Plaçuela de S. Luis.	29. Puerta de la Vega.	36. Corpus Christi.	43. Casa de las Angeres.
2. Carcel de la Corte.	9. la Trinidad.	16. M.° de N.ª S.ª del Carmen.	23. S. Basilio.	30. S. Lazaro.	37. S. Iuste P.	44. A. Comendaderas.
3. Concepçion Geronimo.	10. Calle de la Cruz.	17. Calle de los prescados.	24. Ospital de los Portugeses.	31. S. Maria P. Mayor.	38. Plaçuela del Conde de Varaxas.	45. S. Salasaar P.
4. el Colegio de la Compañia de IHS.	11. Calle de las Carretas.	18. S. Martin.	25. Capuynas.	32. Palacio del duque de Yseda.	39. S. Michel.	46. Descalvas del Carmen.
5. Confesçion Franc.	12. N.ª S.ª de buen succeso.	19. S. Gines P.	26. Carrera de S. Pablo.	33. Los Cannos Viejos.	40. S. Pedro.	47. el Colegio de los Yngleses.
6. S. Millan R.	13. M.° de la Vitoria.	20. Calle del Arenal.	27. la Cruz del espiritu Santo.	34. M.° de las Monjas del Sacramento.	41. Plaçce del duque de Jnfantado.	48. M. de Monjas depinto.
7. la Passion.	14. N. S. de la ynclusa.	21. Hospital de los Franceses.	28. las Beatas Carmelitas.	35. Palacio del Cardinal de Toledo.	42. S. Andries P.	49. Hosp. de los Italianos.

COS DE ESPANNA

MADRID IN 1622, COPY BY EMILIO DE LA CERDA IN 1889 This is the earliest complete map of Madrid, which only became capital of Spain in 1561, having had about 3,000 inhabitants just half a century earlier. Originally thought to have been drawn in the Netherlands, it is now believed the map was drawn by Juan Gómez de Mora (architect of the new Plaza Mayor in 1617–19), coloured by Antonio Marceli and finally engraved by Frederic de Wit, from a family of Dutch printers, in around 1635. This overview was produced using a scale of 1: 6000. The key at the bottom lists the main churches, monasteries, convents, palaces, buildings, streets, plazas and fountains. One detail that assists with the date is that the Palacio de Uceda (key item 32), which stands just above the old Moorish quarter of the city, the Morería, and a short distance west of Plaza Mayor on the corner of Calle Mayor and Calle de Bailén, was built between 1613 and 1625. LEFT

de S. Cathalina.
spiritu Santo.
Antonio.
contra Ansea.
a de S. Domingo.
nesia.
 las Angeles.

5 7 . S. Catalina.
5 8 . Plaça de Horrads.
5 9 . Sant yago P.
6 0 . P. Conde loma.
6 1 . S. Clara.
6 2 . S. Yan.
6 3 . S. Gil Frailes discalsos Franciscanos.

6 4 . el encarnatón Monasterio Real de monjas.
6 5 . Lauadero.
6 6 . Canones del Peral.
6 7 . S. Nicolas.

AREA OF SAN JUAN, MADRID, BY PEDRO TEIXEIRA ALBERNAZ, 1656 This plan, known as the *Topography of the town of Madrid*, was made in 20 large folios that combined into a six-foot by nine-and-a-half-foot map. Created by Teixeira, of the prominent Portuguese cartography dynasty, it is the most important and detailed map of a city in the Habsburg world, produced when Madrid was redefining itself as an imperial city. The plan has a military level of precision, which is reflected in the representation of the buildings with accurate facades, roofs and with the layouts of the courtyard spaces behind the street facades, as shown in this detail from the centre of the city. LEFT

Црⷭтвоуⷲющиⷢ граⷣа Моⷵквⷡ наⷱало гⷪраⷣ всеⷯ Мⷪⷵкоⷡⷡⷥ Рⷭⷣⷡⷵⷡⷣ

КАЛӰСКАН 11.

КОЛӰСКАН

серⷱпⷢуховьскⷶⷤ 10.

Moskua fluuius

Neglina fluuius

петⷬоⷵки W. 1.

ЮБлацки 2.

НИКИЦКИ 3.

ТВⷷСКИ 4.

Дмитⷬоⷵки 5.

петⷬоⷵки 6.

Neglina fluuius

12.

остретⷷⷵки И 7.

покⷬоⷵки 8.

9.
Ивⷪⷵки

Benevole Lec
quadripartita
quarum intima K
Huic proximè adiu
KREMLEN
lâ materiâ adiectâ
cingit TZAR
lapide cingitur, sea
SKORODV
pars huius Meridie
KA SLABON
Magni Domini Cæs

In KIT
numer

1. Troyts; Templ
etiam Hierusalₐ
Palmarum festo, à Cæsa
insidens, à Cæsa

2. Turris cymbala

3. Nalobne meest
culum è latere e
Patriarcha dieb
nonnullos canit
vit publicis prom

4. Plosset, planiti

5. Porta Neglinæ quo

6. Porta ad flume

7. Officinæ ocreas v

8. Tamosene. Telon
ces que importan

9. Mercatorum tab
ces venduntur

10. Tabernæ pictoru

11. Hospitium, quo R
tibus urbibus hos
merces suas vena

12. Officina monetar

13. Aula Anglorum,

14. Vosnesenie. Templ
cuius turris tegum

1. Cæsaris equile

2. Porta ad aquas
at stabulis

3. Hortus herbifer F

4. Nova civium Cu

5. Nosocomium

1. Tzortoffskie

2. Orbaetskie

3. Nikitskie

4. Tverskie

Duodecima porta est i
nona in ligneo muro B

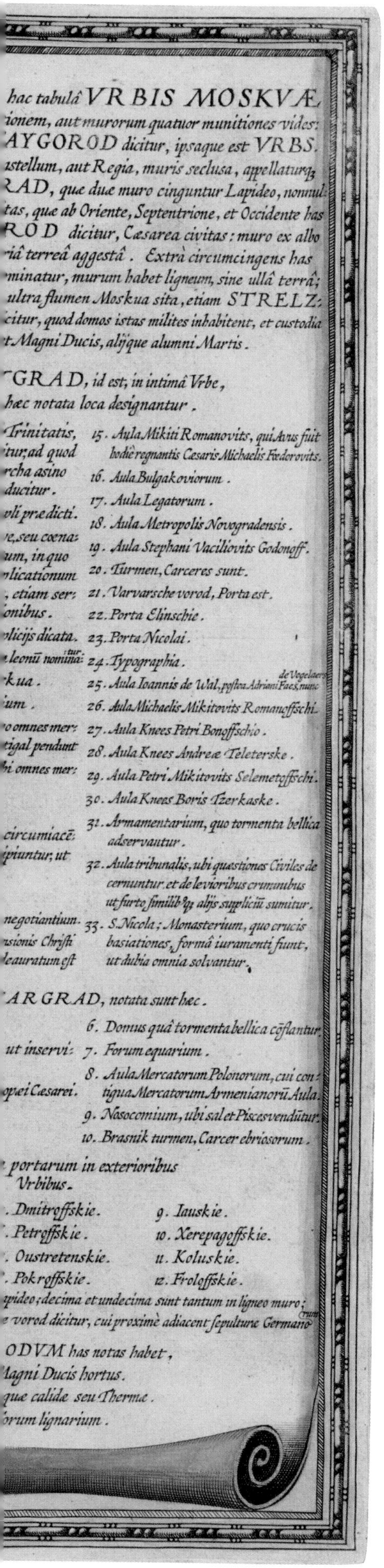

...hac tabulâ VRBIS MOSKVÆ
...ionem, aut murorum quatuor munitiones vides:
...AYGOROD dicitur, ipsaque est VRBS.
...stellum, aut Regia, muris seclusa, appellatur�q̃
...RAD, quæ duæ muro cinguntur Lapideo, nonnul-
...tas, quæ ab Oriente, Septentrione, et Occidente has
...ROD dicitur, Cæsarea civitas: muro ex albo
...riâ terreâ aggestâ. Extrà circumcingens has
...minatur, murum habet ligneum, sine ullâ terrâ;
...ultra flumen Moskua sita, etiam STRELZ:
...citur, quod domos istas milites inhabitent, et custodia
...t Magni Ducis, alijque alumni Martis.

...GRAD, id est, in intimâ Vrbe,
...hæc notata loca designantur.

...Trinitatis,	*15. Aula Mikiti Romanovits, qui Avus fuit*
...itur, ad quod	*hodiè regnantis Cæsaris Michaelis Fæderovits.*
...rcha asino	*16. Aula Bulgakoviorum.*
...ducitur.	*17. Aula Legatorum.*
...uli prædicti.	*18. Aula Metropolis Novogradensis.*
...ve, seu cœna-	*19. Aula Stephani Vaciliovits Godonoff.*
...um, in quo	*20. Turmen, Carceres sunt.*
...licationum	*21. Varvarsche vorod, Porta est.*
...etiam ser:	*22. Porta Elinschie.*
...onibus.	*23. Porta Nicolai.*
...olicijs dicata.	*24. Typographia.*
...leonū nomina:	*de Vogelaer.*
...k.ua.	*25. Aula Ioannis de Wal, postea Adriani Faes, nunc*
...ium.	*26. Aula Michaelis Mikitovits Romanoffschi.*
...o omnes mer:	*27. Aula Knees Petri Bongoffschio.*
...tigal pendunt	*28. Aula Knees Andreæ Teleterske.*
...hi omnes mer:	*29. Aula Petri Mikitovits Selemetoffschi.*
	30. Aula Knees Boris Tzerkaske.
...circumiacē.	*31. Armamentarium, quo tormenta bellica*
...piuntur, ut	*adservantur.*
	32. Aula tribunalis, ubi quæstiones Civiles de-
	cernuntur. et de levioribus criminibus
	ut furto ſimilib'ᵱ alijs suppliciū sumitur.
...negotiantium.	*33. S. Nicola: Monasterium, quo crucis*
...sionis Christi	*basiationes, formâ iuramenti fiunt,*
...deauratum est	*ut dubia omnia solvantur.*

...AR GRAD, notata sunt hæc.

	6. Domus quâ tormenta bellica côflantur.
...ut inservi:	*7. Forum equarium.*
	8. Aula Mercatorum Polonorum, cui con:
...opæi Cæsarei.	*tigua Mercatorum Armenianorū Aula.*
	9. Nosocomium, ubi sal et Pisces vendūtur.
	10. Brasnik turmen, Carcer ebriosorum.

...portarum in exterioribus
Vrbibus.

...Dmitroffskie.	*9. Iauskie.*
...Petroffskie.	*10. Xerepagoffskie.*
...Oustretenskie.	*11. Koluskie.*
...Pokroffskie.	*12. Froloffskie.*

...apideo: decima et undecima sunt tantum in ligneo muro;
...vored dicitur, cui proximè adiacent sepulturæ Germanő

...ODVM has notas habet,
...Magni Ducis hortus.
...quæ calidæ seu Thermæ.
...orum lignarium.

MAP OF MOSCOW, 1662 The *Castellum Urbis Moskvae* is widely considered to be the best mid-17th century plan of the city – the capital of Russia until 1713 – to appear in a commercial atlas. The plan, with south on the left of the orientation, is thought to derive from a survey ordered by the tsar of Moscovy, Boris Godunov (reigned 1598–1605). Amsterdam mapmaker Blaeu's plan shows that the city's defences have multiple levels, which result in several distinctive sections. The royal, religious and secular heart of the city is the inner walled Kremlin (from *kreml* or 'high town', a term that dates from the 14th century) and its later, connecting, Kitai Gorod, or Kitaygorod, meaning 'fortified city', the walls of which were built in 1534–1538.

To the east, north and west is Tsargorod ('the city of the tsar'), protected by white stone walls (leading to the alternative name of Belgorod, or 'white city'), with earth added. These walls are five miles long and have 28 towers, built in 1583–93.

Surrounding that is a wooden palisaded area called Skorodum. The parts of that which lie across the Moskva River are an area called Strelzka Slaboda, which is where soldiers live and senior officials. Finally, there is an outermost earthworks and wooden wall with 12 major gates, constructed in 1591 and known as Zemlyanoigorod ('earthen city'). **LEFT**

AMSTERDAM: The canal ring city

A city built by and for merchants, Amsterdam's commerce and affluence swelled its population from the 1600s onwards. The civic authorities met this challenge by modifying the city layout with carefully planned extensions beyond its perimeter of medieval moats-cum-canals. The creation of a canal ring around three major canals, with hundreds of bridges, has earned Amsterdam the epithet 'Venice of the North'.

AMSTERDAM BIRD'S-EYE VIEW, 1544, BY CORNELIS ANTHONISZ This coloured print of a fine woodcut made from a painting by Anthonisz, in around 1538, when the city had a population of about 12,000, is the oldest surviving plan of Amsterdam. The city is bordered to the north by the bay of the Amstel known as the IJ, which forms the waterfront; to the west is the Singel canal (1428); to the east are Geldersekade and Kloveniersburgwal (1425); and running southwards through the heart is the busy waterway known (later) as Damrak, in the centre of which is Dam. Cargo vessels dropped anchor beyond the city palisade in the IJ and transferred their goods onto lighters, which carried them to Dam to be weighed and sold.

The detail and precision of Anthonisz's work heralded the city's rise to global prominence as a cartographic centre of excellence. Several familiar landmarks can be identified: Oude Kirk (Old Church) in the east, built in 1306, and Nieuwe Kirk (New Church) in the west, built in 1395; on the northeastern edge of the waterfront is Schreierstoren, to the south (inland) of which (at the connection of Geldersekad and Kloveniersburgwal canals) can be seen the former city gate of Het Waaggebouw (modern-day Nieuwmarkt area). At the city's eastern edge is the defensive tower of Montelbaanstoren, with nearby roperies, sawmills and shipyard repairs. **RIGHT**

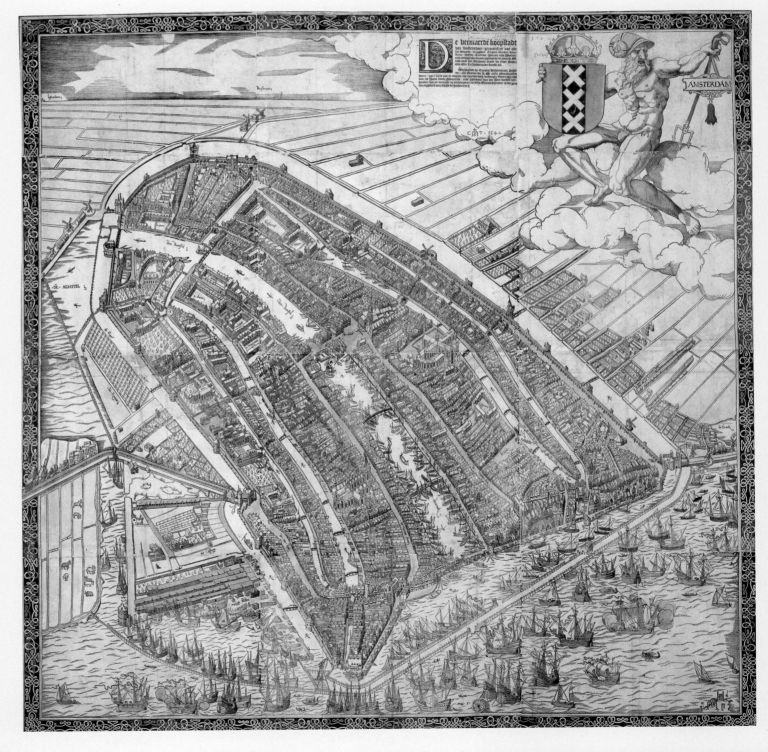

AMSTELODAMI CELEBERRIMI HOLLANDIÆ EMPORII DELINEATIO NOVA.

EXACTISSIMA AMSTELODAMI VETERIS ET NOVISSIMI DELINEATIO PER F. DE WIT

AMSTERDAM, 1649, BY JOAN BLAEU In 1613 the city expanded westwards, beyond the Singel, beginning the distinctive horseshoe ring of canals that exists today. By the time Blaeu produced this map in his *Toonneel der Steden* (*Theatre of Towns*), the three new canals of Herengracht, Keizersgracht and Prinsengracht had been built. Note (at centre right) the monumental new Protestant church, the Westerkerk (built 1620–1631), which stands near the Achterhuis (1635, now Ann Frank House), with the working-class Jordaan district beyond. **ABOVE**

AMSTERDAM, C.1690, BY FREDERICK DE WIT From 1663 to 1672 Amsterdam was expanded eastwards; in combination with the earlier westwards expansion, the development of these new canal districts was a product of town planning and civic wealth as well as expertise in hydraulics and civil engineering. It established the 17th-century model for the artificial 'port city' while giving to the world the distinctive Dutch architecture of façades and gables. **LEFT**

VIEWS IN AND AROUND KYOTO, C.1616–1624 This six-panel screen, in ink, colour and gold leaf, offers a bird's-eye view of Kyoto near the imperial palace during the Gion festival held each summer. The activity depicted here may be on the then important central throughfare of Muromachi–dori, visible running north–south at the right of Kaempfer's map. This belongs to a genre of panoramic cityscapes originating in the city and known as *Rakuchu rakugai zu*, or 'Views in and around Kyoto'. The screen is one of a pair; this one shows the central section of the city and the other the eastern section in the area around the Nijo and Katsura palaces. **ABOVE**

MAP OF KYOTO, C.1690 This map is from *The History of Japan* (1727) by Engelbert Kaempfer (1651–1716), a German physician who worked at the Deshima trading post in 1690–2 before returning to Europe and writing his history. Kyoto, a historic centre near the ancient capital of Nara, was laid out on a Chinese-style grid according to geomantic principles which meant that the mountains to the northeast and northwest offered spiritual protection. The city's eastern and western boundaries were set by the Kamo and Katura rivers, which joined the Yodo River in the south. Kyoto declined in importance with the rise of Edo (Tokyo). **RIGHT**

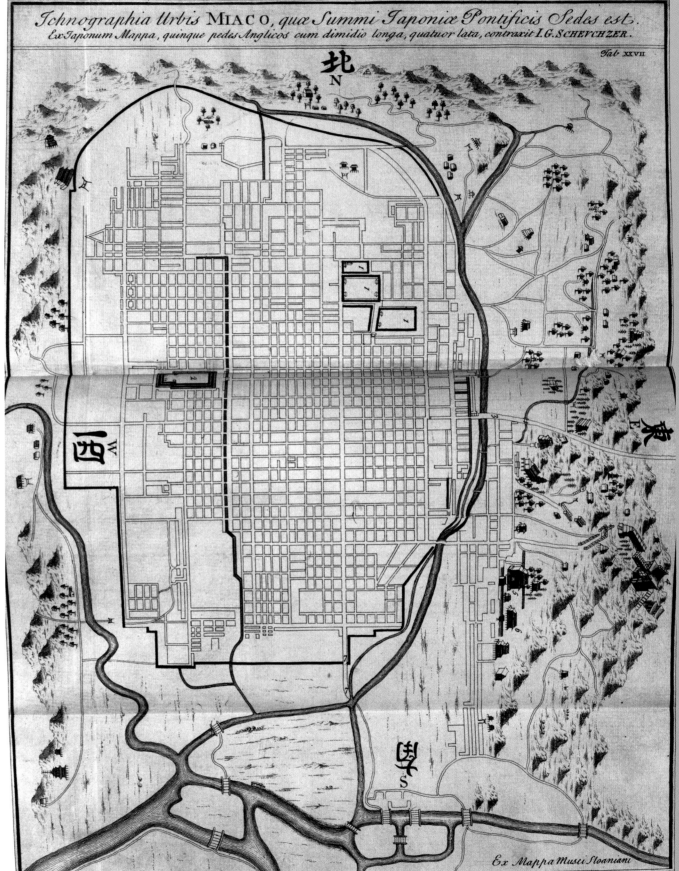

Ichnographia Urbis MIACO, *quæ Summi Japoniæ Pontificis Sedes est.*
Ex Japonum Mappa, quinque pedes Anglicos cum dimidio longa, quatuor lata, contraxit I.G. SCHEVCHZER.

Tab XXVII

北
N

西
W

東
E

南
S

Ex Mappa Musei Sloaniani

Metropolis

MAP OF EDO (TOKYO) C. 1690, BY ENGELBERT KAEMPFER Under the Tokugawa shogunate (1603–1867), Edo was transformed from a fishing village on a marshy estuary in 1590 to the nation's capital by 1650. The city was badly damaged by the Great Meireki Fire of 1657, which destroyed two-thirds. Taken from Kaempfer's *History of Japan*, the central section of this map (north is at the top) shows Edo's moated castle, where Dutch emissaries from the Nagasaki trading post came to pay tribute to the shogun every four years. The emblems, known as *mon*, in the margins represent noble houses, whose leaders had to pay ceremonial visits every other year (known as *sankin kotai*, or 'alternative attendance'), while their wives and children resided there permanently, hostages to their clan's loyalty. **RIGHT**

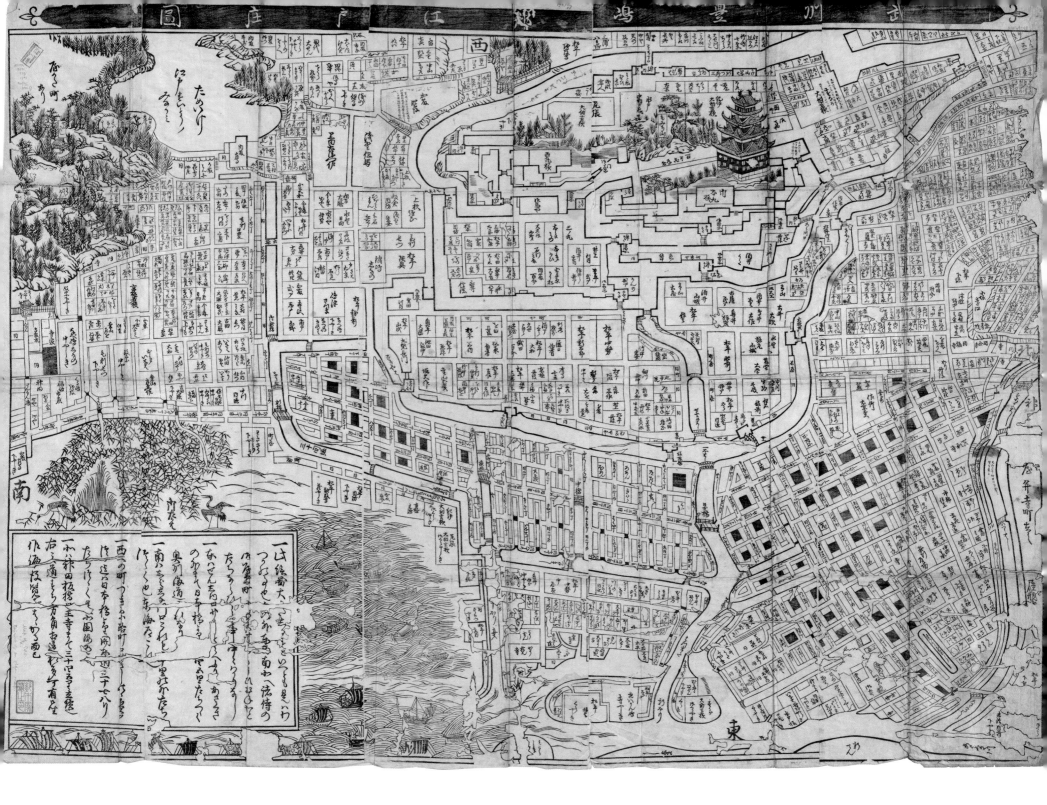

EDO, 1682 This woodblock print cadastral map, with west at the top and Mount Momji depicted pictorially to the south, shows land ownership in central Edo. It is a reprint of a 1630 or 1631 map, which predates the city's fire. The shogun Ieyasu Tokugawa built Edo to legitimise his rule. Most such cities in East Asia followed a grid centred on a north–south oriented palace, but Edo's focus – as the capital of a military ruler – was its vast moated castle. One of the shogun's first acts was to build the large canals and waterways visible in the map: Dosan-bori (1590–1592) linked the bay to the castle, while the Koishikawa and Kanda waterways (1590–1629) fed a large reservoir. Successive shoguns developed the urban layout in a clockwise spiral that radiated from the castle. The *bukechi* residential districts in the vicinity of the castle were for samurai lords, while the more distant districts were for temple authorities or *chouninchi* (commoners). ABOVE

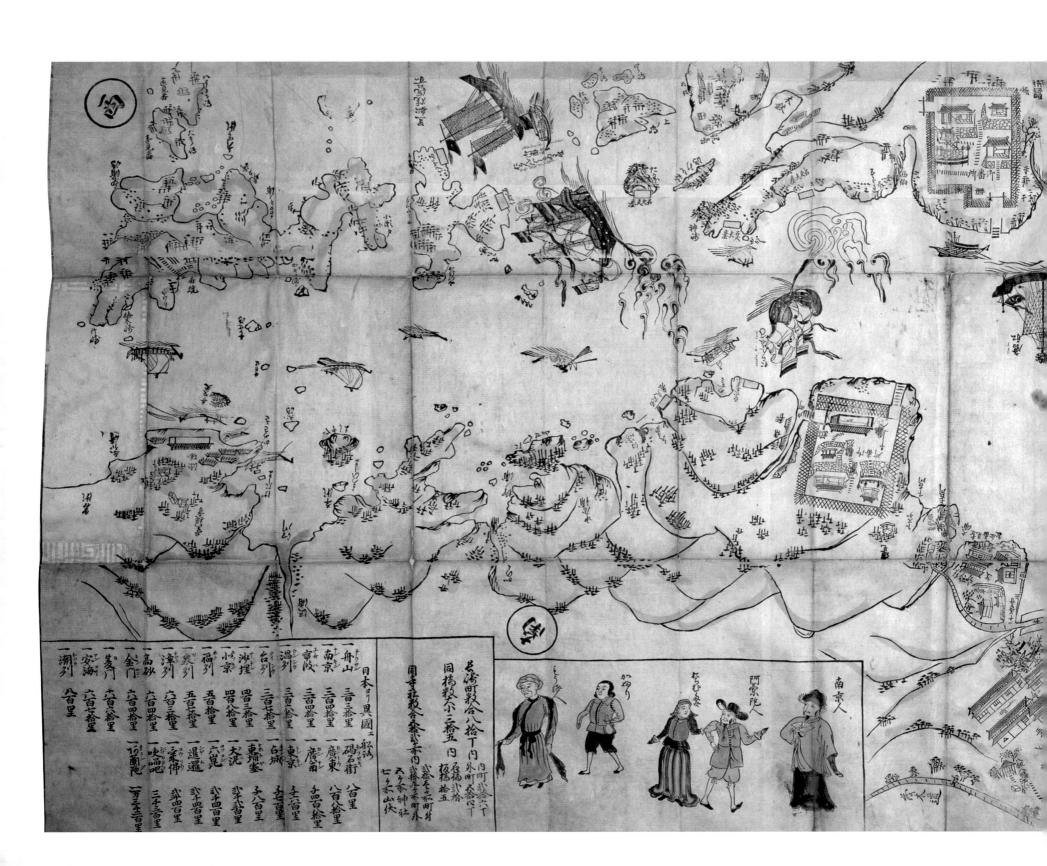

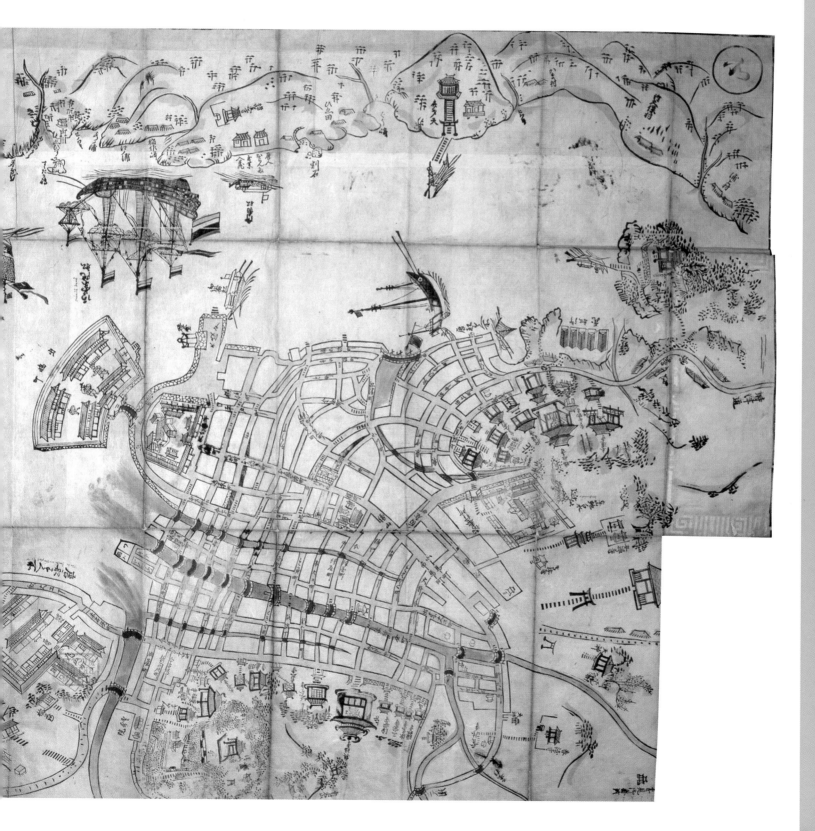

NAGASAKI HARBOUR, WOODCUT ORIGINALLY PRODUCED IN 1680 Following the arrival of the initial Portuguese trading mission from Goa in 1542, the port of Nagasaki on the southern island of Kyushu was opened to European trade in 1570 as the Japanese sought to respond to – and try to control – the arrival of Westerners. In the 17th century, there was a reaction against links with the West, which was accentuated in 1638 when an uprising of Christian converts at Shimabara, near Nagasaki, was suppressed. As a consequence, Christianity was forbidden and the Portuguese were expelled from Nagasaki. The rival Dutch, who had been active in Japan from 1609, managed to win exemption from the ban, as a non-Catholic power, but were confined to the man-made, fan-shaped island of Dejima ('exit island', or Deshima) in Nagasaki harbour in an attempt to manage the trade links. There they conducted business from 1641 until the 1850s. The map is an overhead projection, with south at the top, that shows the entire harbour – the Dutch trading station on Dejima, linked to the mainland by a (guarded) wooden drawbridge, is visible offshore near the centre, to the left of the yellow-coloured street grid. To the bottom left is a Table of Distances from Japan, with various nationalities depicted, including the Dutch and Chinese, who were later given a traders' island to the south of Deshima. **LEFT**

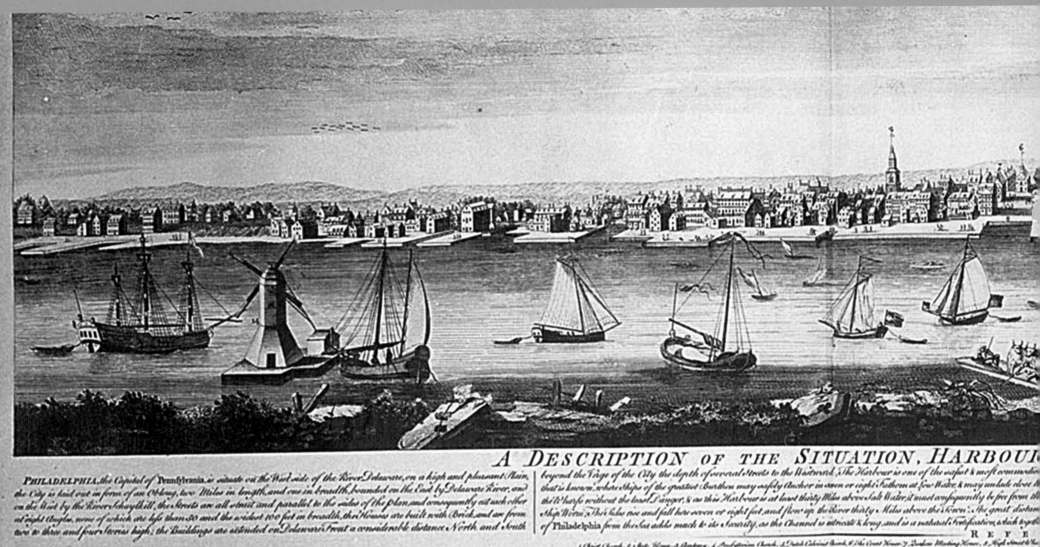

A DESCRIPTION of the SITUATION, HARBOUR

PHILADELPHIA, the Capital of Pennsylvania, is situate on the West side of the River Delaware, on a high and pleasant Plain, the City is laid out in form of an Oblong, two Miles in length, and one in breadth, bounded on the East by Delaware River, and on the West by the River Schuylkill, the Streets are all strait and parallel to the sides of the plan, and consequently cut each other at right Angles, none of which are less than 50 and the widest 100 feet in breadth, the Houses are built with Brick, and are from two to three and four Stories high; the Buildings are extended on Delaware Front a considerable distance North and South

beyond the Verge of the City the depth of several Streets to the Westward, The Harbour is one of the safest & most commodious that is known, where Ships of the greatest Burthen may safely Anchor in seven or eight Fathom at Low Water, & may unlade close to their Wharfs without the least Danger, & as this Harbour is at least thirty Miles above Salt Water, it must consequently be free from the Ship Worm; The tides rise and fall here seven or eight feet, and flow up the River thirty Miles above the Town; the great distance of Philadelphia from the Sea adds much to its Security, as the Channel is intricate & long, and is a natural Fortification, which together

REFE

1. Christ Church. 2. State House. 3. Bank. 4. Presbyterian Church. 5. Dutch Calvinist Church. 6. The Court House. 7. Quakers Meeting House. 8. High Street

PART OF SCHULKILL RIVER

DELAWARE RIVER

Vine Street

Mulberry Street

High Street

Chestnut Street

State House

The Corporation Burying Ground

Christ Church

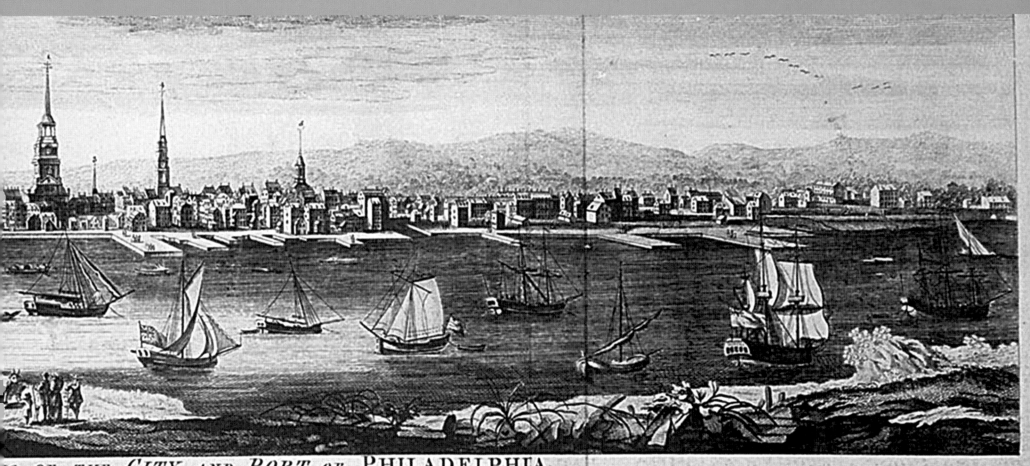

C. OF THE CITY AND PORT OF PHILADELPHIA.

THE BATTERY

AN EAST PROSPECT OF PHILADELPHIA, C.1768, BY GEORGE HEAP AND NICHOLAS SCULL This engraving, produced by Thomas Jefferys in London, was one of the most important colonial-era city views and it is based on a view of the city from across the Delaware River that was recorded by George Heap under the supervision of Nicholas Skull, Surveyor General of Pennsylvania.

The plan shows the city waterfront and the harbour of the port, busy with many ships in the foreground, accompanied by a plan of the city between the Schuylkill and Delaware rivers, and vignettes of the Battery and the State House. The view of the city extends from present-day South Street to Vine Street and shows considerable detail. The steeple of the State House, or Independence Hall, is visible at the left, while Christ Church is recognisable in the centre. The accompanying key identifies 14 places of interest, including the Quaker Meeting House and the Presbyterian Church. PREVIOUS PAGES

Although Europe was increasingly important in the urban world, China and India still dominated the roll of leading cities. The population in East Asia was greater than that in Europe. Leading Chinese commercial centres, notably Nanjing, Hangzhou and Suzhou in the so-called Jiangnan area of the lower Yangze delta, were major cities, as was Beijing, the imperial capital from 1421. In South Asia, by the early 1700s there were British footholds in India at Calcutta, Madras and Bombay (Mumbai), and in 1799 the Mysore capital of Seringapatam was captured. However, other major cities were not yet under British control or protection, notably Delhi. In Southeast Asia, the fall of capital cities were key episodes, as in 1752 when the Mons of Pegu captured Ava, the capital of Upper Burma. The Burmese then invaded Siam (Thailand), attacking the capital Ayuthia, but to no lasting effect.

URBAN AND ECONOMIC GROWTH

Of the 19 cities in the world in 1800 believed to have had a population of over 300,000, only five were in Europe: London (third), Constantinople [Istanbul] (eighth), Paris (ninth), Naples (fourteenth) and St Petersburg (seventeenth). Other cities were still significant, however, both for their own societies and, in some cases, as centres of an expanding global trading system. Of the 1.1 million inhabitants of Lombardy in the 1700s, 130,000 lived in the capital, Milan. In 1755, the 260,000 inhabitants of Lisbon, capital of an empire, constituted about 10 per cent of Portugal's population. Amsterdam's population was nearly twice that of the Dutch provinces of either Friesland or Overijssel.

More generally, it is difficult to link urbanisation to economic activity or growth. Large cities were as much a feature of relatively stagnant Mediterranean Europe (Constantinople and Naples) as of growing northwestern Europe (Amsterdam and London).

Maps do not necessarily record prosperity or economic roles. For example, maps offered little on the role of cities in the slave trade even though they were the key points. Cities such as Cartagena in Spain and Veracruz in the Gulf of Mexico were major ports for landing slaves, while in Lima (Peru), Mexico City (Mexico) and Havana (Cuba) slaves were employed as craftsmen, labourers and servants, notably in the construction and maintenance of Havana's fortifications. The need for finance, especially in the sugar commission business, ensured that London, where such moneylending was centred, was ultimately as heavily involved in the slave trade as a thriving port such as Liverpool.

Atlantic maritime commerce more generally was the major factor behind the growth of some traditional Mediterranean ports, notably Marseille in southern France, which grew enormously as a result of the West Indies trade, which accounted for a quarter of activity in the port before the end of the century.

PLACES OF POWER

Many urban activities had only a tenuous relationship with economic growth. For example, like London, the growth of Paris was characterised by an expansion westwards, with many fine houses built for the wealthy, and in the case of Paris these were for the nobility, notably in the Faubourg St-Germain. Cities were points of power, and displays for this power, as with major urban palaces and government buildings. A renewed development of the age was the desire for governments and rulers to demonstrate their power, vision and sense of modernity by creating new urban forms, sometimes where nothing had existed previously. In Germany Karlsruhe, by Karl Wilhelm,

THE PALACE OF THE MARGRAVE OF BADEN-DURLACH IN KARLSRUHE, 1739, AFTER JOHANN JACOB BAUMEISTER Royal and princely palaces were often significant regional and national structures. Versailles, Schönbrunn and Nymphenburg were located in the countryside, but others (Stockholm, begun in 1697; Berlin, begun in 1701) were built in the cities. The avenues of the town of Versailles converged on the nearby palace, and a similar effect was achieved for Karl Wilhelm's wall-less dream city of Karlsruhe, begun in 1715.

The new *residenzstadt* was intended to demonstrate the principles of geometric town planning, with the palace the focus for 32 streets radiating fan-like and arranged within a concentric boulevard. The nine streets to the south of the palace made up the 'new town', which was bisected by the wide Langestrasse as the major commercial thoroughfare. **LEFT**

margrave of Baden Durlach, and in Russia St Petersburg, by Tsar Peter the Great, both showed how cities minor and major could be built from new. Imposing squares also played a role in this trend, as with the Kongens Nytorv and Amalienborg Slotsplad in Copenhagen.

Cities were centres of consumption and service industries as much as of commercial and industrial activity. The map trade was an aspect of both categories. London was its prime centre, a process encouraged by its commercial wealth and contacts, and by the entrepreneurial freedom of the city.

Nevertheless, Amsterdam and Paris remained of cartographic importantance. Moreover, maps were produced in other cities, including Hamburg and Nuremberg.

Cities were also a major topic for cartography. London was served by John Rocque's new survey in the 1740s, but it was not only the major cities that were mapped. James Corbridge's map of Newcastle (1723) offered both precision and an attractive image, illustrating the map with pictograms of the major buildings, for example the leading churches. Isaac Taylor followed the same policy in his map of Wolverhampton (1750) and such architectural motifs became commonplace.

NEW CITIES FOR THE NEW WORLD

The developing world of Europe's colonies ensured expansion for the Western model of cities. Cities were founded as British North America expanded, including Baltimore in 1728 and Richmond in 1730. These foundations had an impact on the urban hierarchy of North America and a more significant consequence on relations within particular colonies. Within Maryland, the rise of Baltimore had an impact on Annapolis, while in Virginia, that of Richmond represented an inland move of the centre of gravity from Williamsburg. The foundation of Savannah in 1733, at once port, capital, centre of population and, crucially, military base, was vital to the anchoring of the new colony of Georgia.

Existing cities also grew. Founded in 1682, Philadelphia's population rose from 2,500 in 1685 to about 25,000 by 1760, making it one of the biggest cities in the English-speaking world with one of the busiest ports in the western hemisphere. Pennsylvania's Proprietor, Thomas Penn, the son of the founder of the colony, sought to attract further inward investment

SURVEY OF THE COUNTRY ON THE BANKS OF THE HUGHLY RIVER EXTENDING FROM THE TOWN OF CALCUTTA TO THE VILLAGE OF OOLOOBAREAH, 1780–84 Published in October 1792, this map was the product of Lieutenant-Colonel Mark Wood's survey of 1784–1785, officially aimed at the 'Convenience of the Health of the Inhabitants of the said Settlement'. His survey yielded a town plan that identified European habitation, and a map that showed the area for three miles around on both banks of the Hooghly.

The first known British map of Calcutta was published in 1742 and the colonial administration was the major patron of cartography for the next 200 years – its interests reflecting the nature of the survey undertaken or the map created. These maps evolved from plans that reflected a military engineering (defensive) perspective to ones concerned with effective urban planning and improvement.

Wood's map shows the Old Fort as the nucleus of the urban settlement (the fort became a customs house after 1766) on the western bank of the Hooghly, with the vast new star-shaped Fort William – set within a massive cleared area known as the Maidan – to the south of the town.

RIGHT

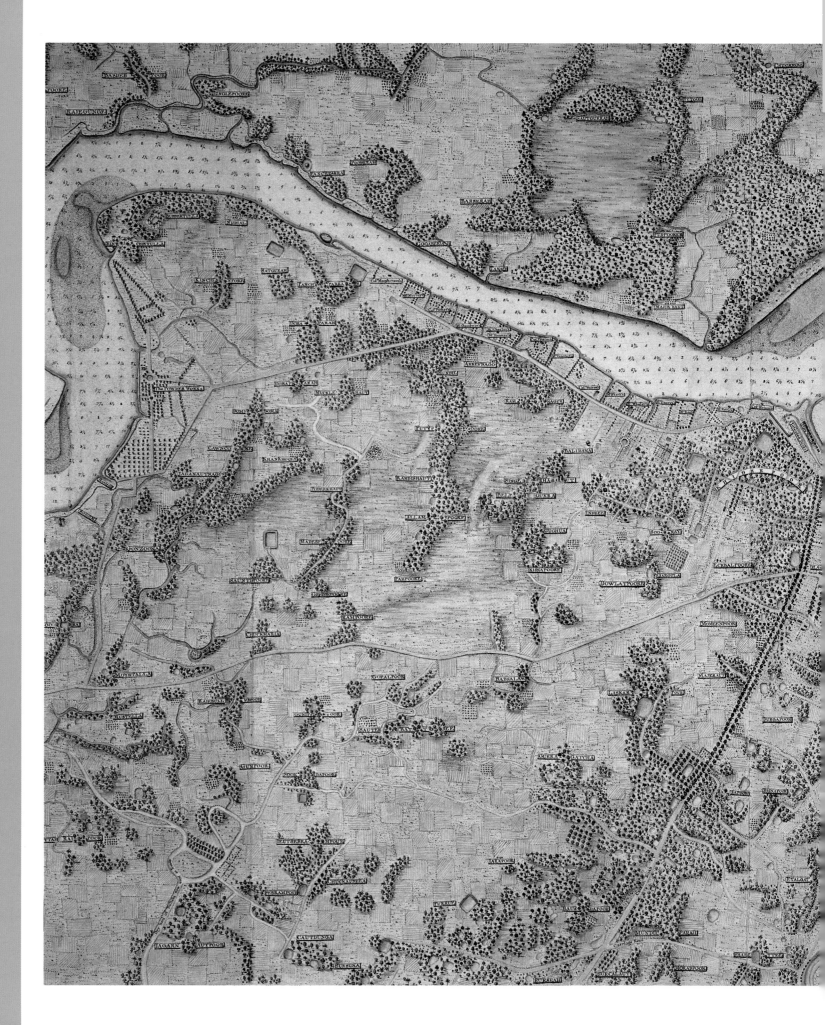

To his most Excellent Majesty
George the 3.rd

THIS SURVEY of the COUNTRY
on the BANKS of the HUGHLY RIVER
Extending from the TOWN of CALCUTTA
to the Village of OOLOOBAREAH,
and which includes the FORTRESS of FORT WILLIAM
and WORKS at BUDGEBUDGE
also Representing the Soundings of the River at Low Water
in Spring Tides
SURVEYED by order of GOVERNMENT 'twixt the YEARS 1780 and 1784
is Most Humbly Presented
by his most Faithful
and devoted servant & subject
Mark Wood
Lt Colonel Chief Engineer

SCALE of 2 MILES or 10560 FEET

by commissioning a large (seven-foot-long) panorama of the city from landowner and businessman George Heap, first issued to much acclaim in 1754.

Cities were the centres of opinion-forming not only due to their often vociferous populations but also their press. In 1690, the unlicensed *Public Occurrences Foreign and Domestic* was published in Boston, but swiftly suppressed. In 1704, the *Boston News-Letter* became the first regular newspaper in British North America.

CENTRES OF TRADE AND COSMOPOLITANISM

New governmental arrangements, for both administration and trade, were significant in the development and ranking of cities. In Spanish America, the viceroyalties of New Granada and the Rio de la Plata (River Plate) were established in 1739 and 1776 respectively, based on Bogota and Buenos Aires. The captaincy-general and presidencia of Caracas were established in 1776 and the *audiencia* of Cuzco in 1787. These new agencies were part of a marked increase in the presence of government. Thus, the new capital of Caracas had a treasury, a high court of justice and a regiment of troops. Moreover, under the Free Trade Decree of 1778, a direct trade between Buenos Aires and Spain began.

Trade helped make cities cosmopolitan, notably in Asia where Westerners were far more dependent on regional commercial networks. In 1772, Dean Mahomet, an Indian in the Bengal army, found that in Calcutta: '... the greatest concourse of English, French, Dutch, Armenians, Abyssinians, and Jews, assemble here; besides merchants, manufacturers, and tradesmen, from the most remote parts of India.' The survey of the city initiated by the Commissioners of Police in 1780 (and carried out in 1784–1785 by Mark Wood) resulted in a map of the town that distinguished between areas 'inhabited by Europeans

PLAN OF LIVERPOOL, 1765, BY JOHN EYES An inset of the city's new Exchange (1749–1754) confirms the growing port of Liverpool's devotion to commerce. The rural areas of Toxteth Park to the east and Shaw's Brow to the north both became densely populated in the next century. A charitable Infirmary opened in 1749 on Shaw's Brow was testament to the civic benefits of encouraging trade. Much of the identifiable industry is related to shipping (roperies, shipyards). The Old Dock (built in 1715) was filled in during the 1820s as the city's docks expanded ninefold from 1756 to 1836 alone to cope with a thirtyfold rise in tonnage. **RIGHT**

PLAN OF THE CAPITAL CITY OF ST PETERSBURG WITH THE DEPICTION OF ITS MOST DISTINCTIVE VIEWS, 1753, BY THE IMPERIAL ACADEMY OF SCIENCES AND ARTS Commissioned to celebrate the 50th anniversary of the building of the city in 1703, 100 copies of this plan were printed and accompanied by 12 city views by the artist Mikhail Makhaev. The attributes of the art of war in the upper right corner betray the city's origins as a fortified city and naval base on the Baltic to challenge the power of Sweden The first structure to be built was Petrapavlosk, the Peter and Paul fortress on an island in the Neva River. To the south is Vasilevskiy Island, connected by pontoon bridge to the Admiralty and Winter Palace on the bank opposite. **OPPOSITE**

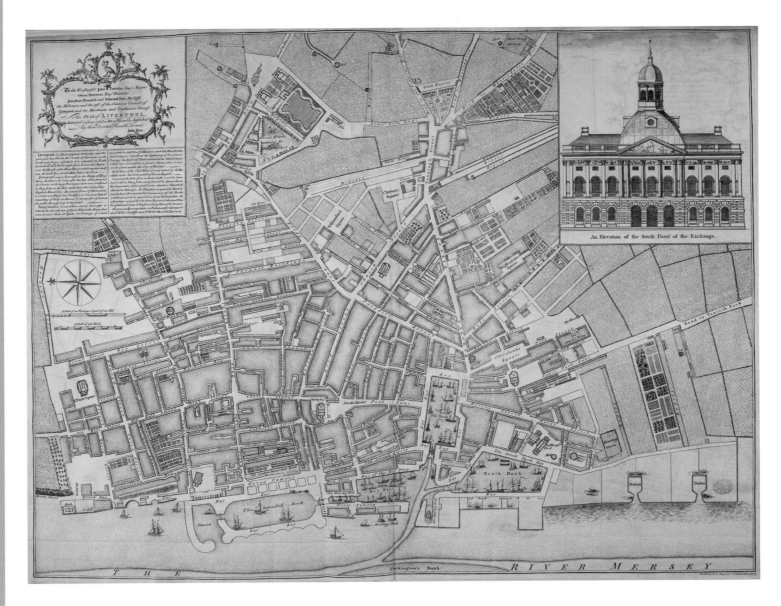

... [and those] ... inhabited by the Natives'. For the first time the streets of Calcutta were named in a map.

Yet the relationship between nationalities was not always good, particularly between the Dutch and Spanish authorities and Chinese immigrant communities. In Batavia, the Chinese were treated harshly from 1722 while they evaded immigration quotas. The situation boiled over in 1740 when Dutch fears of a rising and Chinese fears of expulsion led to a crisis that became a massacre in which 10,000 Chinese were slaughtered. The crucial role of the Chinese in the economy of Batavia was such that, in subsequent censuses in 1778 and the 1810s, they comprised a large percentage of the population. There was a similar massacre in Manila in 1763.

MAPS AND REALITIES

The technology of map production did not really change much during the period. However, the image of the city altered. In place of perspective views, scaled town-plans became dominant.

It is instructive to compare the impression created by maps with the records of tourists, although the former could not really conjure up the atmosphere or capture the character. Paris was described by David Mallet in 1735 as 'that metropolis of dress and debauchery'. Sarah Bentham in Rome in 1793 found the churches magnificent, but: 'I was much disappointed in seeing Rome. The streets are narrow, dirty and filthy. Even the palaces are a mixture of dirt and finery and intermixed with wretched mean houses. The largest open places in Rome are used for the sale of vegetables. The fountains are the only singular beauties.'

Maps could at least capture a particular situation, matching views in so doing. Tourists and other visitors frequently commented on the contrast between situation and interior, as when Frederick, second Viscount Bolingbroke, wrote harshly of the city of Naples in 1753: 'The town is very ugly but finely situated.'

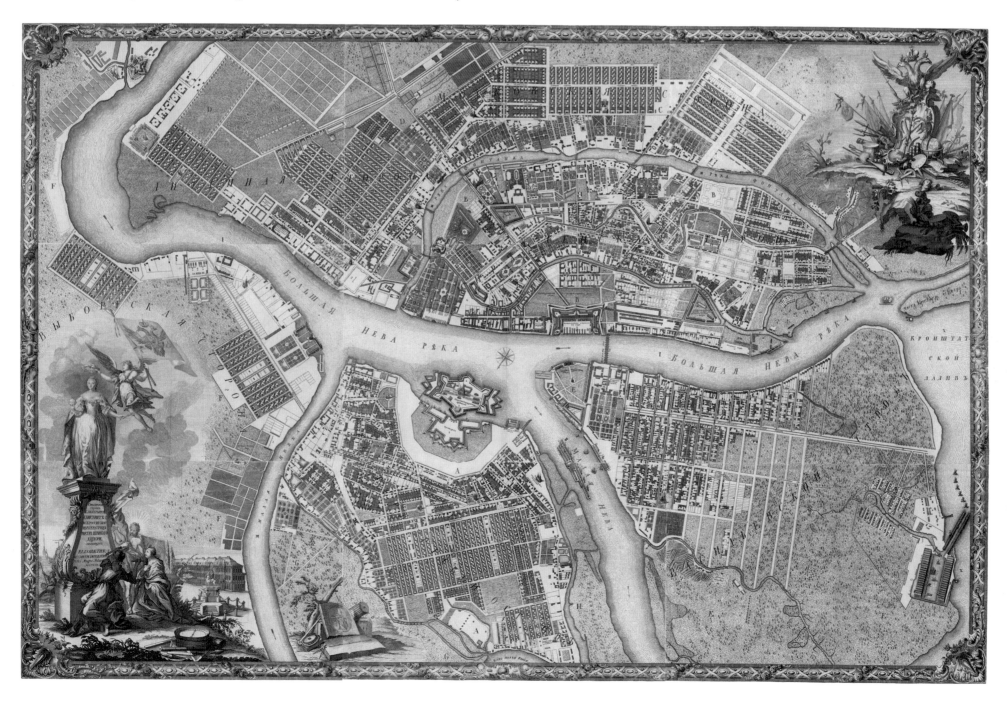

SAVANNAH, 29 MARCH 1734, BY PETER GORDON
Founded in 1733 by General James
Oglethorpe (his tent,2, is pitched near the
stairs, 1), the new colony was laid out on a
bluff overlooking the Savannah River. The
layout uses a basic urban neighbourhood
unit of square-based blocks, known as
'wards', approximately 600 feet north–
south and 540–600 feet east–west. Each
ward was bounded by streets, facilitating
good transport links but on a pedestrian
friendly scale. An open square had four
residential blocks and four civic blocks
(each divided by east–west lanes). The
plan allowed for the logical expansion of
the settlement over time, and endowed
the city with a great deal of open space.

The map shows the first four squares
taking shape. In the southwest quadrant a
square (the edge of its open area marked
by two pairs of trees) has buildings at
three of the cardinal points (5, 6 and 10; 9
is unbuilt) in what will be its civic blocks.
In the blocks to the north and south,
most of the houses are built. During the
American Revolution, Savannah was
captured by the British in 1778 and
unsuccessfully beseiged by a French-
American army in 1779. **RIGHT**

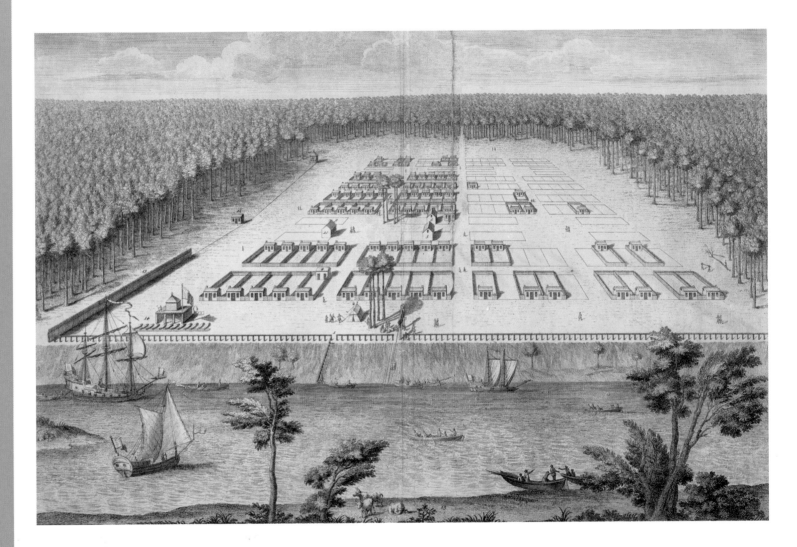

MONTRÉAL, 1734, BY JOSEPH-GASPARD CHAUSSEGROS DE LÈRY The French colonial
city was founded in 1642 as an outpost for trade and the military, known
then as Ville-Marie. By 1734 (when a fire destroyed a quarter of the city's
houses), some 2,000 people lived within its nearly completed city walls,
building of which had begun in 1717, under the guidance of military
engineer Joseph-Gaspard Chaussegros de Lèry.

Although Montréal's walls were removed in 1804–17, some early
structures still survive, making it one of North America's oldest urban
spaces. The Sulpician seminary (1684) on Place d'Armes is the city's
oldest building. South of Place d'Armes is Pointe-à-Callière, where this
once dense urban mosaic first emerged. **RIGHT**

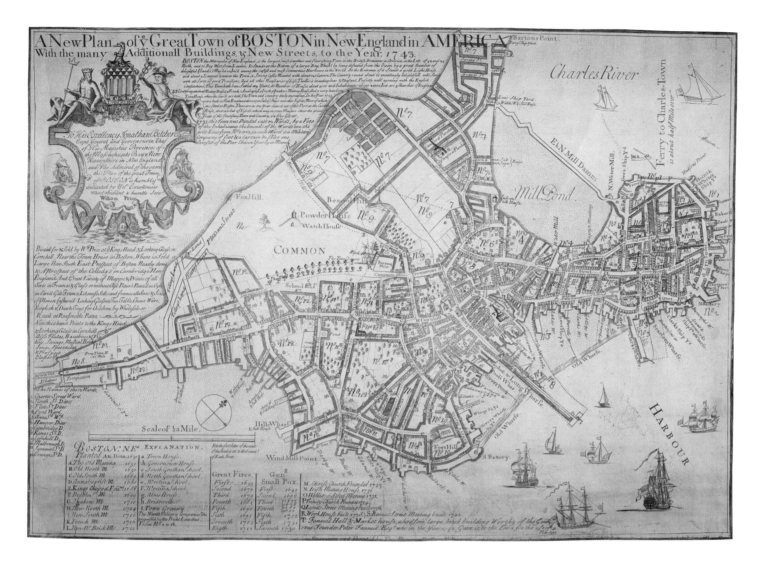

A NEW PLAN OF YE GREAT TOWN OF BOSTON, 1743, BY WILLIAM PRICE Oriented northwest, and based on an earlier (1722) work by John Bonner, this map bears some fascinating detail by virtue of its indexes (bottom left) of wards and significant buildings, plus dates of 'Great Fires' and epidemics of 'Small Pox'. The Common and the so-called Mill Pond dominate the topography. At this point in its history the city was suffering from a period of economic stagnation, but the citizenry was hostile to many attempts to relieve this by building markets. In 1740 merchant Peter Faneuil's proposal to build, at his own expense, a new market on the site of a failed development atop the recently filled-in Town Dock was passed narrowly in a 367–360 vote. By 1742 Faneuil Hall was built and appears (indexed T) as a 'hadsom large brick building'. **LEFT**

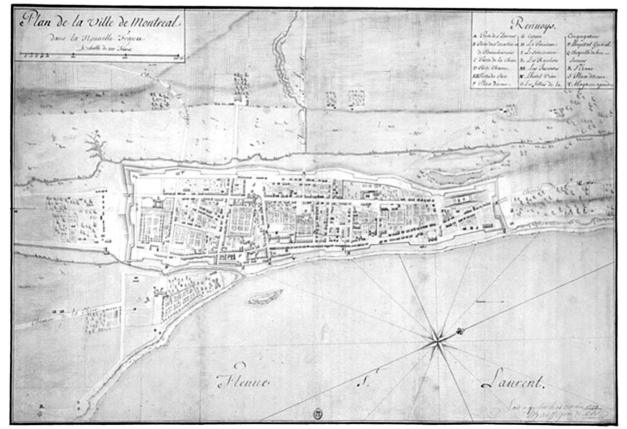

**DIFFERENT VIEWS OF THE MAJOR CITIES IN PERSIA,
1762, BY JOHANN BAPTIST HOMANN** Despite
the title, this print shows aerial views of
15 cities from the Caucasus to
Afghanistan, each with a key to the most
important sights. From top to bottom
and left to right, the four rows show:
Astrakhan and Derbent in Russia, Tbilisi in
Georgia and Kars in Turkey; Erzurum in
Turkey, Baku in Azerbaijan, Sultanieh in
Iran and Shamakhi in Azerbaijan; Yerevan
in Armenia, Shiraz in Iran, Kandahar in
Afghanistan and Ardabil in Iran; and
Kashan, Isfahan and Bandar Abass, Iran.
OPPOSITE

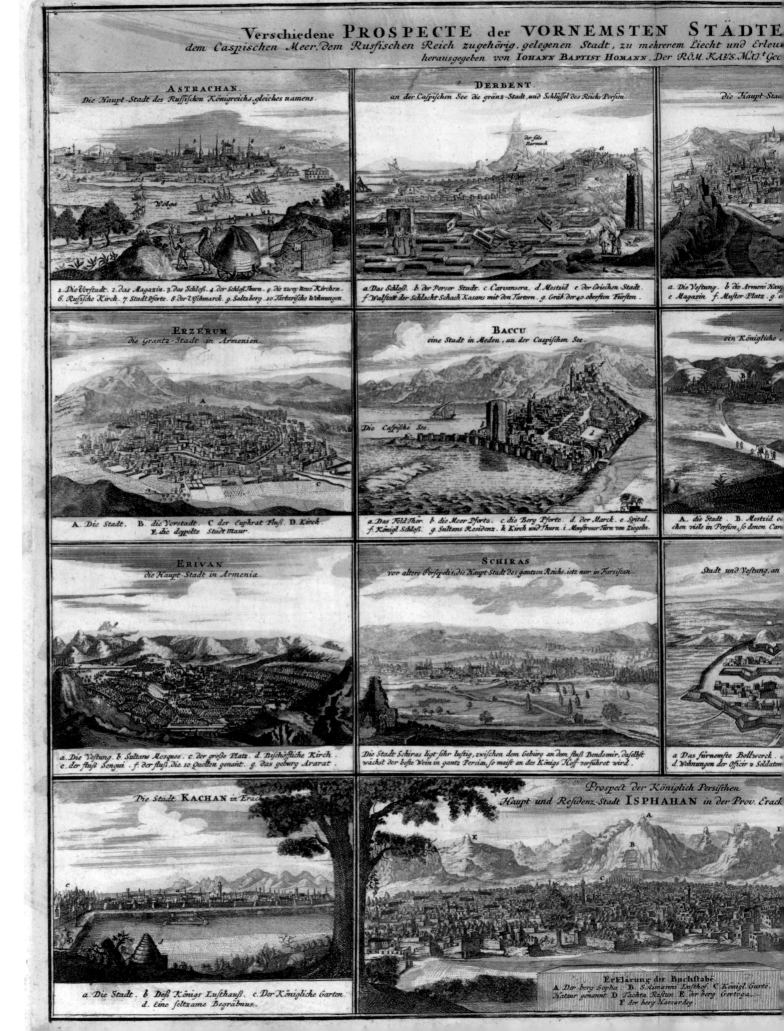

in **PERSIEN** samt vorderst einer unfern
g der neu-verfertigten Persianischen Land-Charten
ho in Nürnberg.

LIS.
rgistan oder Georgien.

KARS.

4. c. Cathobische Kirch. d. Capuciner Closter.
rck. h. zum H. Creutz. i des Printzen Palast.

A. *Das Schloß.* B. *die Stadt.* C. *Armenische Kirch.* D. *der Kars fl.*
E. *Weg nach Erivan, so 9. Tag. reisen von Kars liget.*

ANIA
n der Prov. Erack Atzem.

SCHAMACHIA
im Gebürg der Provincz   Schirwan.

ianische Kirch. C. eine Karvansera, derglei.
zur sichern Einkehr als Gasthöfe dienen.

A. *Deß Chans Residenz-Schloß und Wohnung.* B. *die innere Stadt.*
C. *die Vorstadt.* D. *die alte ruinirte Vestung.*

AHAR
ianischen Gräntze des Reichs Persien.

Die Stadt ARDEBIL, *in Adirbeitzan.*

ndere Bollwerck. c. das Gouverneurs Hauß.
arck. f 2 Dämme, worauf man zur Stadt gehet.

A. *Schich Sefi und der Könige begräbnus.* B. *Mestzid Adine, die Haupt kirch.* C. *der grosse Marck-platz.* D. *der Saltz Marck.* E. *deß Chans Wohnung.* F. *das Müntz Hauß.*

rthia.

GAMRON *oder*   BENDER ABASSI

1. Die Holländische Wohnung.   2. die Englische.   3. die Französische.
4. Mestzid oder Moschée.   5. Das Castell.   6. das Fort.   7. der berg Ginau.

PLAN OF THE FORT AND TOWN ON THE PROMONTORY OF GOOD HOPE, 1750, BY JACQUES-NICOLAS BELLIN Founded by the Dutch in 1652, Cape Town was used as a base on the route to and from the Dutch East Indies, and it was the Dutch East India Company (VOC) that built the star-shaped fort in 1666–1679. The nearby town grew steadily during the 1700s and was laid out on a grid radiating from a central square. A tree-lined canal ran from the VOC's gardens, through the town alongside the hospital, around the grand parade area and into the sea. By the time the French cartographer Bellin mapped it there were more than a thousand fine houses in close proximity to the shipyards, taverns and warehouses along the shoreline, which catered for the dozens of ships and crews which laid anchor in the bay each year. Cape Colony became a slave society, but it was far less developed than the plantation economies of the New World. **RIGHT**

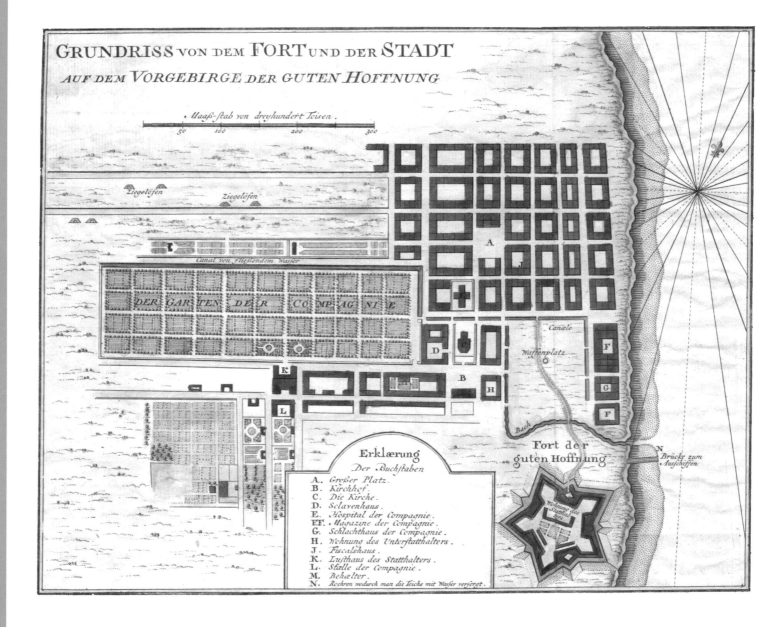

GRUNDRISS VON DEM FORT UND DER STADT AUF DEM VORGEBIRGE DER GUTEN HOFFNUNG

Maaß-stab von dreyhundert Toisen.

Erklærung
Der Buchstaben

A. Großer Platz.
B. Kirchhof.
C. Die Kirche.
D. Sclavenhaus.
E. Hospital der Compagnie.
EF. Magazine der Compagnie.
G. Schlachthaus der Compagnie.
H. Wohnung des Unterstatthalters.
J. Fiscalshaus.
K. Lusthaus des Statthalters.
L. Stelle der Compagnie.
M. Behælter.
N. Roehren wodurch man die Teiche mit Wasser versorget.

PLAN OF KINGSTON, JAMAICA, BY MICHAEL HAY, C.1745 1737–52 This cadastral map has the distinction of being based on one of the most detailed plans of any British colonial city. Produced by surveyor Michael Hay for Edward Trelawny, the governor of Jamaica, it depicts the island's new principal settlement during the golden era of the sugar economy. Kingston was founded as Jamaica's commercial centre in 1693 to replace the nearby city of Port Royal, which had been destroyed by an earthquake. Many of the first settlers were refugees from Port Royal. Hay's survey is based on a layout by John Goffe and military engineer Christian Lilly in 1702. They devised Kingston as a parallelogram with a rational grid of streets, at the centre of which was a four-acre square, known as Parade (used initially as the military camp and where there was a water supply), at the junction of King and Queen streets. The grid covered 240 acres, and was half a mile from east to west, and a mile from north to south (North Street to Port Royal Street on the waterfront). Each property lot in Hay's map is identified with a number that corresponded to its owner, while the upper sides of the map show prominent homes, which had been built in a signature Jamaican style. **RIGHT**

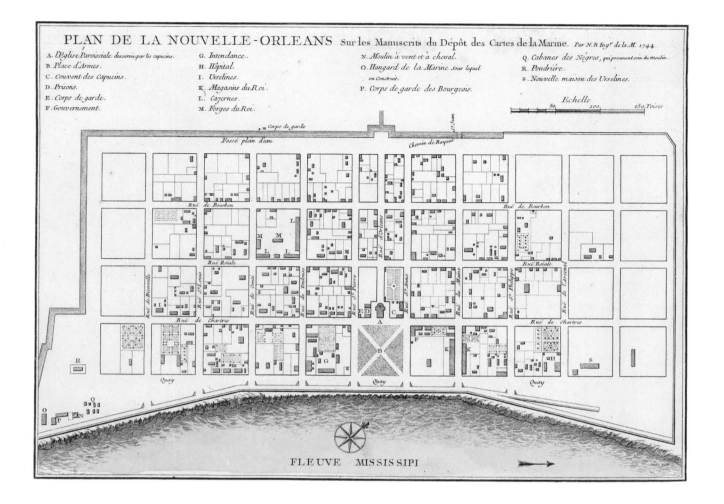

PLAN OF NEW ORLEANS, 1744, BY JACQUES-NICOLAS BELLIN Showing approximately 100 buildings, 18 of them identified in a key, this is the earliest map of the city founded by the French in 1718 along the Mississippi River and named after Philip, Duke of Orléans, regent of France for the infant Louis XV. Although, earlier French bases had been founded in Biloxi Bay (1699) and at Mobile (1702), New Orleans became capital of Louisiana.

Oriented to the east, the map shows the original settlement inland as far as the marked *fossé plein d'eau* ('ditch full of water'), corresponding with modern-day Dauphine Street. The grid radiates from riverfront Place d'Armes (now Jackson Square), the heart of the area now known as the French Quarter, or Vieux Carré. **LEFT**

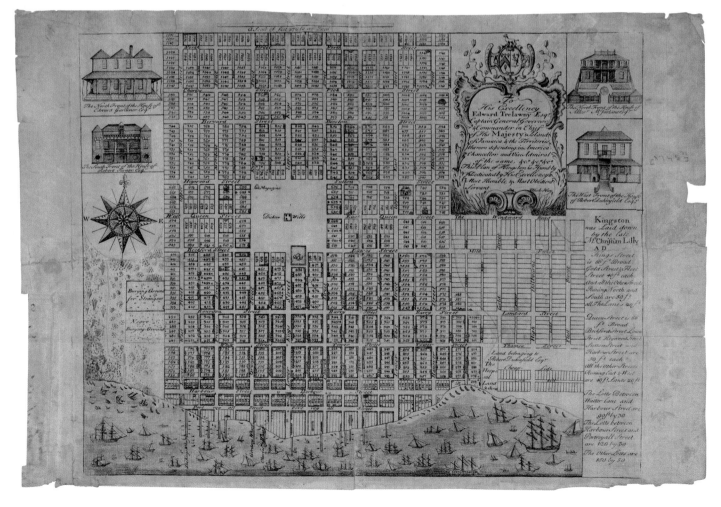

Metropolis

LONDON, BY JOHN ROCQUE, 1746 *The Survey of London Westminster & Southwark*, engraved by Rocque's partner John Pine, is the most accurate and detailed representation of the 18th-century metropolis that exists, with some 5,500 street and place names. This is not to say that it is absolutely accurate – for example, churches are scrupulously recorded but industrial buildings are not. Financed by subscription, Swiss-born Rocque began his surveying in 1738, the methodology of which included trigonometrical surveying to build up a lattice of points measured between church steeples. Differences between those measurements and his detailed ground survey slowed down the project, which was not available until 1747.

The 'Plan' – Rocque described it as such rather than a 'map view' – has a scale of 26 inches to the mile and it is most accurate for the most densely developed parts of the city, such as the West End. Many smaller streets and courts are omitted while 'considerable Houses and Gardens' are included. One contemporary detail not left out was the 'Tyburn Tree' – the gibbet for public executions on the edge of 'Hide [Hyde] Park' at 'Tiburn' (Tyburn), near where Marble Arch stands today. South London includes Southwark's ancient settlements near London Bridge as well as dockyards in Deptford and the area round Lambeth.

RIGHT

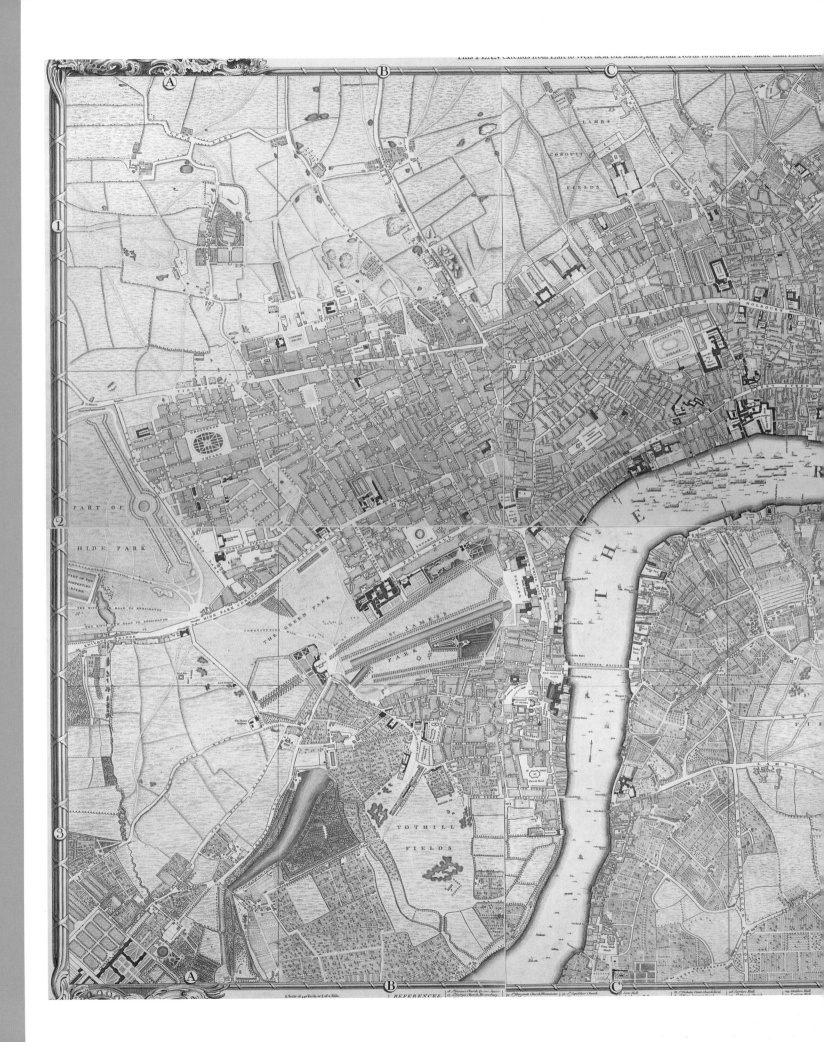

NEW YORK, 1766–1767, BY BERNARD RATZER This magnificent city plan is the most detailed and finely engraved map of any city of the 13 American colonies before the American Revolution. New York City, which then occupied only the tip of Manhattan Island, is depicted in exacting fashion, delineating all streets and the major buildings and diverse places of worship, while the topography of the fields and forests surrounding the city is captured with great care. Across the East River are the farms of Brookland (Brooklyn) that supplied the city with their agricultural bounty by using the ferry (marked on the map).

The map was based on one produced, during the Stamp Act riots in the winter of 1765, by the military engineer John Montresor, which was printed in London in 1766. The governor of New York, Sir Henry Moore, commissioned Ratzer (misspelled as 'Ratzen' on the map) to make a more precise survey, while using Montresor's plan as a base. ▶

Ratzer's map reveals who owned the farms and large estates, how the city's infrastructure began to develop from a network of roads, where the natural features lay, and how the landscape had begun to be changed by the creation of pastures and gardens. The map was not actually printed until 1770, in London. When war broke out in 1775, New York City was soon seized by the Revolutionaries, but in 1776 the British drove them out. A series of mysterious fires in lower Manhattan then destroyed one-third of the city, which meant that by September 1776 the New York that Bernard Ratzer had recorded just a few years before no longer existed. **LEFT**

EDINBURGH: city of two towns

The capital city of Scotland was awarded World Heritage Site status in 1995, because of its urban architectural heritage. In a uniquely characterful way, Edinburgh combines a well-preserved ancient city, or Old Town, which has medieval high-rise buildings, narrow streets and public buildings, with a well-planned, 18th-century neoclassical addition, or New Town, where fine Georgian residential terraced homes were built for many of the leading citizens of the day – prominent individuals who transformed the city in the era known as the Scottish Enlightenment, when it earned, for intellectual and aesthetic reasons, the name of the 'Athens of the North'.

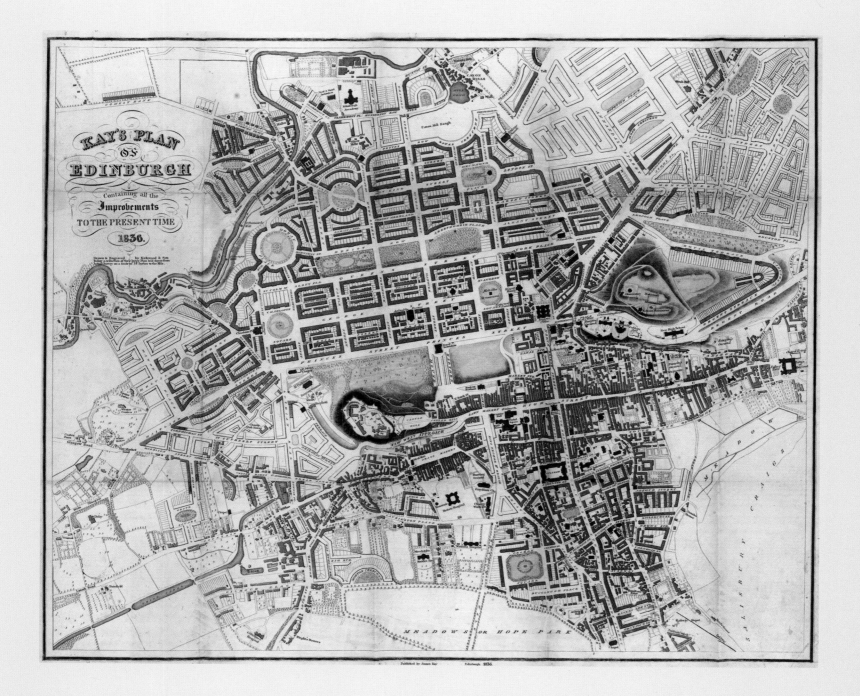

EDENBOURG (EDINBURGH), 1582 BY BRAUN AND HOGENBERG Shown in plan view from the south, with figures of nobles dressed in the manner of the era of Mary Queen of Scots, this map is based on an earlier woodcut by Raphael Holinshead in 1574. The west is dominated by the castle, indicated as *Castrum puellarum* ('Maidens' Castle'), possibly because princesses were able to live there in safety. It is connnected by the long horizontal thoroughfare known as the High Street or 'Royal Road' (later the Royal Mile), which leads via the church of St Giles to the town gate and Holyrood abbey. At this time the city had around 15,000 inhabitants, many of whom lived in densely inhabited, insanitary, medieval multistorey buildings with as many as 12 floors. **LEFT**

EDINBURGH, 1836, BY JAMES KAY The first plan for the layout of the New Town in Edinburgh was won in a competition by James Craig, whose formal and orderly proposal of three main streets was adopted in 1767. By the late 1790s the plan had largely been put into effect, involving significant contributions by figures such as Robert Adam. In the 1800s a series of extensions to the New Town were undertaken, enlarging it northwards, eastwards and westwards. Kay's map of 'improvements' made to the city reveals the contrast between the orderliness of the new and the chaos of the old. The map also shows how the transition between the two towns was effected by draining the boggy Nor' Loch area and creating Princes Street Gardens, with a connecting mound formed out of earth resulting from the New Town excavations. **ABOVE**

SKETCH MAP OF THE NEW CAPITAL DISTRICT, BY THOMAS JEFFERSON, 1791 After securing its Independence from Britain, the United States did not want its capital city to be Philadelphia, the seat of the Continental Congress. A compromise between the northern and southern states resulted in a site on the Potomac River. On 31 March 1791 Thomas Jefferson sketched these ideas for the new federal district: the areas of Georgetown and Rock Creek are marked at the top left, and the Mall is annotated with 'President, public walks, Capitol'. Jefferson believed that not only were wide open streets aesthetically pleasing, but would also be better for public health. The new city was built according to a 1791 grid plan by engineer and architect Pierre Charles L'Enfant that focused on the Capitol. RIGHT

PIERRE CHARLES L'ENFANT'S PLAN OF WASHINGTON, 1792, BY ANDREW ELLICOTT This map, by the original surveyor of Washington, DC, presents a simplified scheme of the new city, without street names or lot numbers. The notation identifies the longitude as 0 degrees, making the new city the prime meridian – an issue dear to Ellicott, who wished to make a clean break from Greenwich which derived from the former colonial power. Ellicott depicts George Town in outline, names two buildings (President's House and Capitol) and shows a bridge spanning what is now the Anacostia River. FAR RIGHT

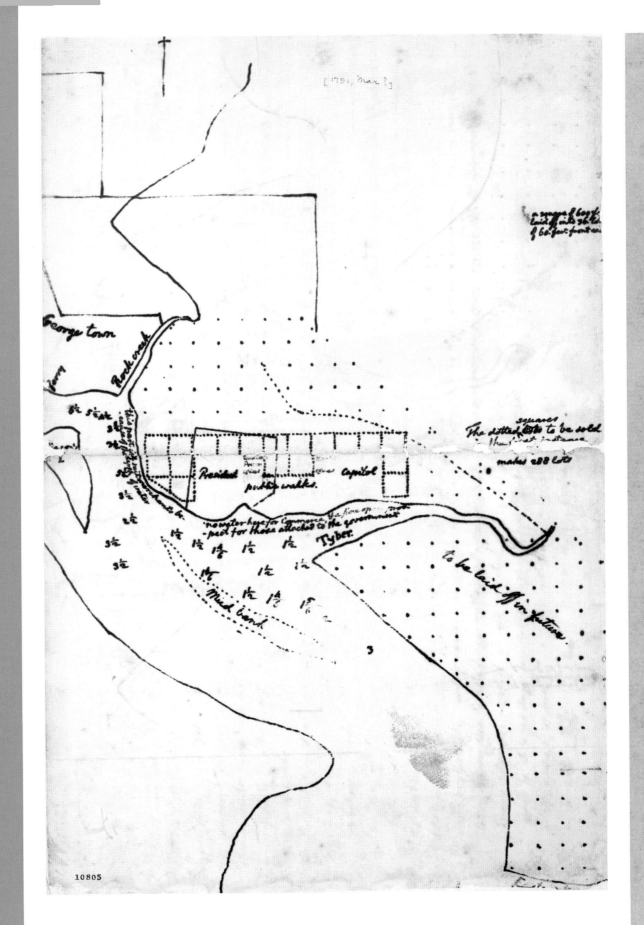

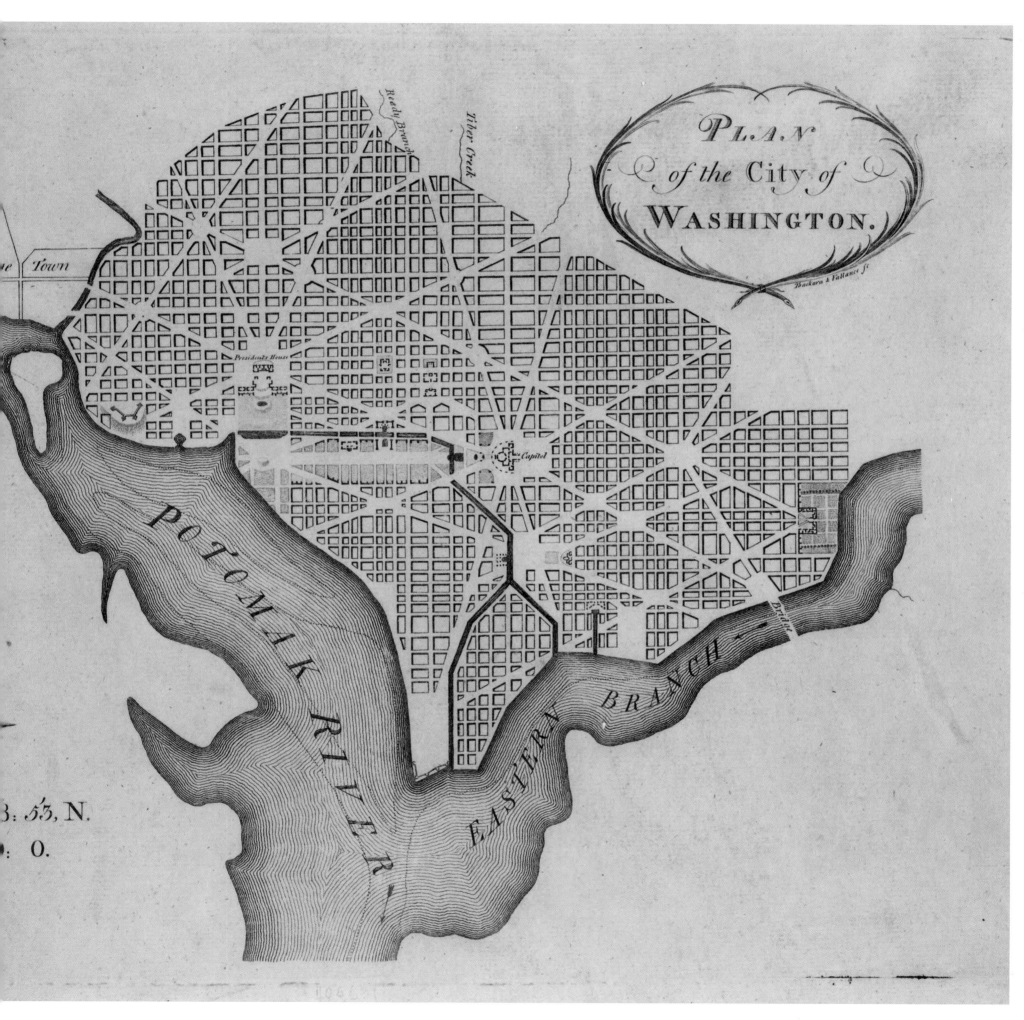

PLAN of the City of WASHINGTON.

Reedy Branch

Tiber Creek

Town

President's House

Capitol

POTOMAK RIVER

EASTERN BRANCH

Bridge

B: 53, N.

0.

THE 1734–1736 MAP OF PARIS, 1739, BY LOUIS BRETEZ Better known as the Turgot Plan, after the city's principal merchant (effectively mayor), Michel-Étienne Turgot who decided to promote Paris by commissioning the greatest example of urban cartography up to that time. Begun in 1734, the detailed measurements took two years, and the resultant bird's-eye view map of the entire city was published in 1739 as an atlas of 20 sections which cover an area that corresponds to 11 of the 20 modern-day arrondissements. Technically, the plan has a scale of 1:400 and is an axonometric projection, oriented towards the southeast, and the level of detail is extraordinary – the inset here is from the centre of the main image and shows the Pont Neuf near the Quai de Conti on the Left Bank. RIGHT

Metropolis

PARIS AND ITS PRINCIPAL BUILDINGS, 1789, BY JACQUES ESNAUTS AND MICHEL RAPILLY First published by M. Pichon in 1784, this map went through several significantly revised editions, the last of which appeared in 1802. Adorned with a masthead featuring a nymph of the Seine and an allegory of Paris with the symbols of the arts and sciences, the map shows Paris immediately prior to the French Revolution. Bisected by the River Seine, Paris is rendered in finely engraved detail and has indexes in each margin that identify important streets, parishes, colleges, hospitals, buildings, squares and other places of interest, which can be pinpointed by using the letter and number reference in the horizontal and vertical axes. Decorating the plan's outer border are views of 27 structures in Paris and the Palace of Versailles. In August 1792 the royal palace, the Tuileries (bottom, centre), was stormed by Revolutionaries. The declaration of a republic followed. **RIGHT**

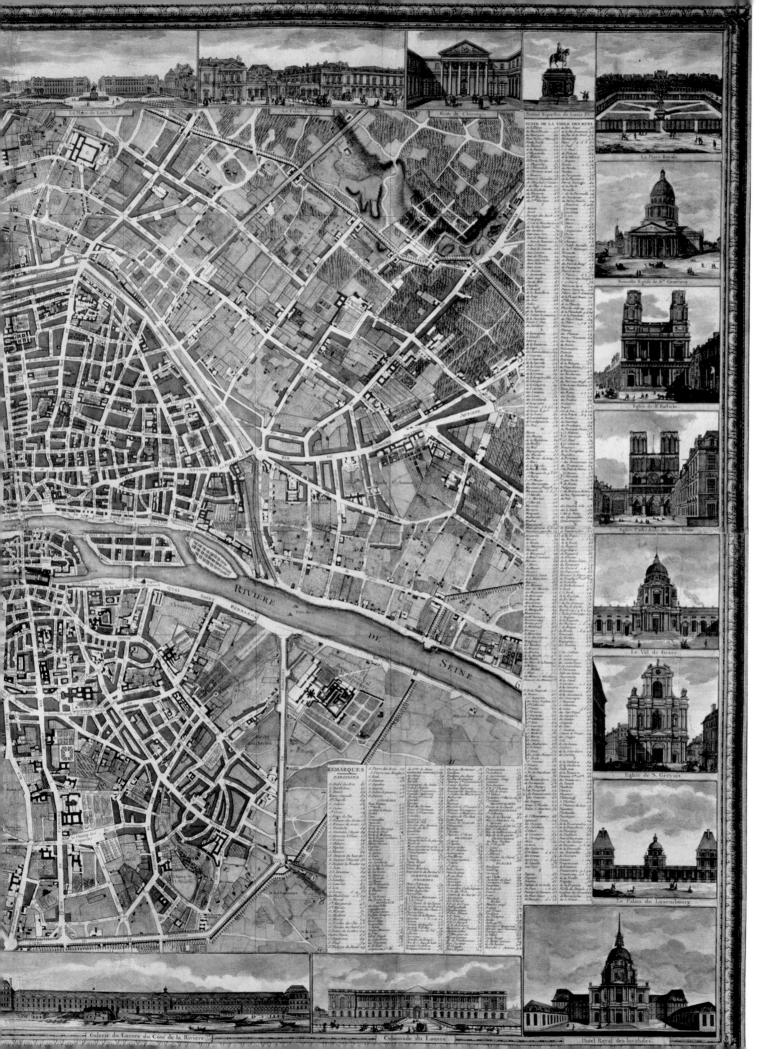

MEXICO CITY PLAN, 1794, BASED ON IGNACIO CASTERA Oriented with east at the top, this pen-and-ink plan by Manuel Ygnacio de Jesus de Aguila is based on a 1782 map produced by Castera for Viceroy Pacheco de Padilla Revillagigedo as part of his scheme for urban improvement. The Indian lands beyond the city limits (such as the cluster in the northern Santiago area) were felt to be disordered concentrations of poverty. The viceroy – using colour coding to indicate the condition of streets and housing – wished to improve drainage, public health and paving, while expanding the city towards the cardinal points where four new focal plazas (public squares) would be developed. In the event, only the paseos (promenades) in the southeast and southwest were incorporated. RIGHT

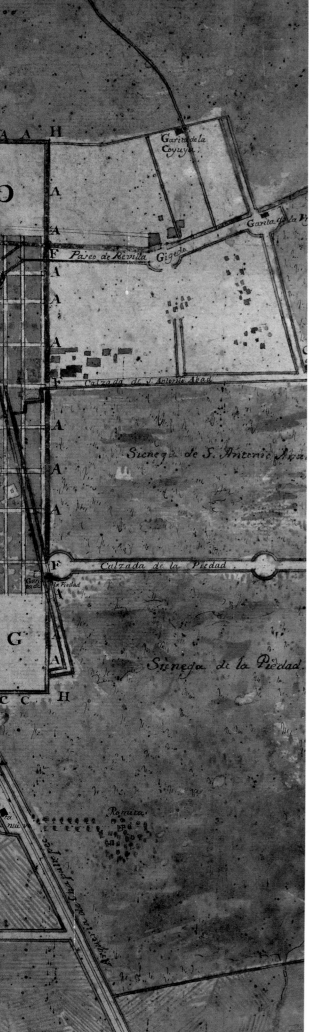

A PLAN OF THE CITY OF MUNICH, PUBLISHED BY J. STOCKDALE, LONDON, 1800 Munich, the well-fortified capital of Bavaria, was fundamentally reshaped between 1806 (when it became a kingdom) and the late 1800s. Stockdale's plan captures the city immediately before the beginning of several phases of modernisation, showing clearly the perimeter belt of Baroque-era fortifications erected under Elector Maximilian I, beyond the earlier medieval walls. The Hofgarten (Court Garden – item 2 in the plan) lay within that outer defensive ring. However, in 1795 Elector Karl Theodor declared that Munich 'is not, cannot be, and should not be a fortress,' and announced his intention to create a new, 'open city' by razing the walls and removing any obstacles to the city's growth and planned expansion.

Although published in London in 1800, the drawing for this map is thought to date from about 1760. It certainly appears to predate 1798 when the Englische Garten (English Garden), the largest city park in Europe, was laid out in 1789–1792 to the area north of the Schwabinger city gate and the Hofgarten. **BELOW**

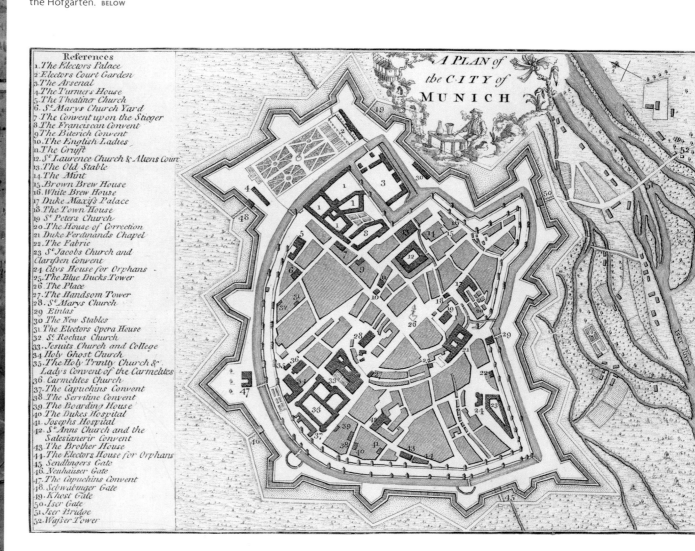

THE RED FORT OF DELHI, BY MAZHAR ALI KHAN, NOVEMBER 1846 This is only a detail from a dazzling, five-panel panorama of Delhi, the northern Indian city where Muslim rule in the subcontinent was established in the 12th century. The picture is based on observations made from beneath a cupola on a city gate and it captures the moment when Delhi's ruling sultanate was dying. Little more than a decade after Khan's masterpiece, the Indian Mutiny of 1857 and its suppression meant that to ensure control the British changed the face of Delhi and its once leafy expanses forever, with mosques, palace buildings and others close to the walls demolished for reasons of security. **PREVIOUS PAGES**

THE JAGANNATH TEMPLE, PURI, 19TH CENTURY, ARTIST UNKNOWN The seaside city of Puri on the Bay of Bengal in Orissa has a local art form known as *patachitra*, which represents a cult form of Krishna for which Puri is the premier place of worship, attracting many pilgrims to visit the city. Known as the Jagannath ('Lord of the World'), the deity is depicted figuratively and spatially within a temple. The *patachitra* is rectangular and depicts the trinity of Jagannath with his brother Balabhadra and sister Subhadra in a sanctum within a walled enclosure. The world outside the temple shows other temple shrines within Puri, then the oceans and heavens beyond the city. **RIGHT**

The 19th century was an age of technological change, imperialism, population growth, industrialisation and urbanisation unprecedented in history. The world's rising population also became more urban: the 1851 Census in Britain recorded that the British were predominantly urban for the first time in their history. Other countries followed, although more slowly. Partly as a result, Britain's cities were key subjects for urban mapping.

INDUSTRIES AND POPULATIONS

A number of developments contributed to this urbanisation, including greater agricultural productivity as well as the importation of food from distant areas, notably Australasia (lamb and mutton), Argentina (beef), and North America (grain). As a result, fewer people needed to live on the land in western Europe and the northeastern USA in order to feed greater numbers of people in the cities. Moreover, the large factories to which industrialisation gave rise employed substantial workforces, and many cities, such as London and New York, were industrial to an extent that is not generally appreciated today. Provincial centres with traditional industries, such as Lyon in France, also grew – in Lyon's case adding chemicals and pharmaceuticals to its continuing strength in textiles, notably silk-weaving. The 1860 census in the USA revealed a population of 31.5 million, a third of whom were supported directly or indirectly by manufacturing industry.

Rapid urban population growth was fuelled by migration because many cities had high death rates (for example, in 1840 in Glasgow, life expectancy at birth was 27). Many people found work not in industry but in commerce, transport, construction and administration, as well as in domestic service. However, incomers were faced with insufficient housing stock, which led to overly high densities of population – a key element of poverty. In 1801, Liverpool had a population of just over 80,000 and it increased more than fourfold within half a century. The city was a key port dominating Britain's Atlantic trade, but it was also faced with terrible overcrowding, poor sanitation and disease. The 19th century was an age of statistics and in Liverpool, which contained some of the country's worst housing, the annual death rate was 34.4 per 1,000, while in 1846 the city contained 538 brothels and in 1857 there were at least 200 regular prostitutes under the age of 12. Alongside the extensive and impressive world-leading docks that were prominent in contemporary maps of the city, these sombre statistics hinted at the human suffering that accompanied the growth in trade.

TRANSPORTATION ARTERIES

Cities were also the centres of transport networks, a prime theme for mapping. In this era, cities owed much of their growth to their ability to shape and profit from new transport systems, which early in the century meant canals and roads, with the former crucial for bulk freight. Thus, London benefited from the establishment of canal links to the English Midlands. In the USA, the opening of the Erie Canal in 1825 provided New York with a major position on the transport system to the West. Maps reflected the extent to which some national urban hierarchies were reshaped by canal routes.

This process was further boosted by the invention of railways, which played an important part in the rise of cities such as Chicago. Meanwhile, despite the transcontinental expansion of the USA to reach the Pacific Ocean, towards the close of the 19th century the country was still dominated economically by the cities of the eastern seaboard from Baltimore to

Boston, albeit with the addition of the abutting area to the west that included Chicago, Milwaukee and St Louis. Apart from their financial and corporate dominance, these cities contained much of the manufacturing activity and population.

There was a feeling in the South and the West that the cities of the East dominated them economically, financially and politically. This tension was the background to the anti-big business, trust-busting measures of the early 20th century. Southern politicians and business interests sought rail links to provide an alternative to northern links focused on Chicago. This was notably so with the Southern Pacific Railroad, which was completed to California from New Orleans on its own lines in 1883.

THE INFRASTRUCTURE OF IMPROVEMENT

Cartographers everywhere were kept busy as the face of the city was transformed by the train – because, unlike canals, railways moved both freight and passengers. The railways stimulated suburbanisation, which altered the shape of cities and towns by enabling many of the newly wealthy middle classes to move into leafy, outlying suburbs. Such changes created a need and a demand for new maps. Moreover, railways needed major new infrastructure, from railyards to bridges to stations. These redefined the inner shape of cities as well as providing a new focus for the transport networks within them, from cabs and buses to the revolutionary new form of mass-transit system: underground railways.

Metropolis

Maps both explained and aided the rapid expansion of cities. In less than a century Houston in Texas went from next to nothing to being one of the largest cities in the USA. An 1849 survey of Los Angeles recorded lots around the city offered to anyone willing to spend $200 on improvements. Maps of the city from the mid-1850s onwards showed that it was surrounded by plains that provided an ideal area for urban sprawl.

At the same time, the political potency of the major centres of population increased with the spread of democratic systems of government, which was the case in most of Europe, the New World and Australasia. In Britain, the Great (First) Reform Act of 1832 increased the significance of cities. Redistribution ensured that major cities got their own MPs. Other countries did likewise and the pace of reform gathered momentum.

Cities were places where ruling powers could be violently overthrown, which happened in several European states in 1848, the Year of Revolutions. Risings in Paris, where royal authority had been overthrown in 1792, brought down the system of government in 1830 and 1848. In 1852, partly in order to make it easier for the state to exert control over the city, as well as to ensure that Paris was an appropriate imperial capital for his empire, Napoleon III had Baron Haussmann construct new boulevards in the 1850s and 1860s, a modernisation programme that was also undertaken elsewhere – albeit with different motivations and solutions – in places such as Copenhagen, Vienna and Berlin.

As well as beautification, many urban 'improvement' schemes were designed to destroy

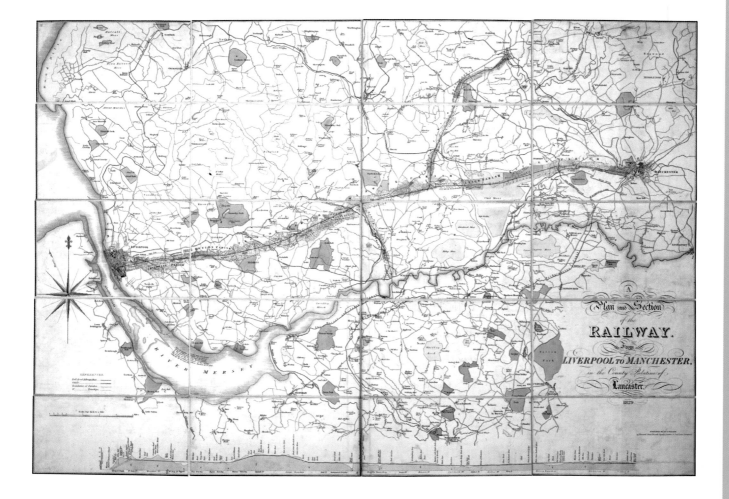

GEORGE STEPHENSON'S MAP OF THE LIVERPOOL AND MANCHESTER RAILWAY, 1827 Opened in 1830, this was the world's first intercity railway, built to ferry goods (such as cotton) to and from Liverpool faster and more economically than the canal system. However, the cut in journey times was quickly seized upon by passengers too. With time, long-distance networks were built and services from London reached Birmingham in 1838 and Southampton in 1840. Within cities, commercial patterns were changed; districts were demolished and reconfigured to accommodate the lines, goods yards and centrally located terminus structures; and urban street layouts were greatly affected by the new focus on railway stations. LEFT

urban slums, known in London as 'rookeries', which the authorities felt were not really under their control and were centres of crime and disease that presented a threat to all citizens. Public health issues in particular played an increasing role in urban planning as the century progressed. Redevelopment linked to social control was also seen in the less successful cities of the era. In Rome, the poor were moved in the 1870s from central areas into shanty towns on the outskirts, while in Naples a cholera epidemic in 1884 was followed by slum-clearance near the port and the building of a major road – Corso Umberto I, named after the king – through the city centre.

CARTOGRAPHY AND TECHNOLOGY

Technological advances in publishing greatly affected mapping by making it possible to create more and less expensive maps. Steam power enabled large-scale production and this meant that maps could be aimed at a mass market. Mechanised papermaking became commercially viable in the 1800s, leading to plentiful quantities of inexpensive paper. The steam-powered

printing press developed in the same period. From the 1820s, lithography made a major impact and contributed to an increase in the quantity and range of maps made. In lithographic transfers, the design for an engraved plate was transferred to a litho stone on which alterations could be made – for example, the redrawing of boundaries – without affecting the original plate. Alterations might be made on the stone between different printings of the map, which made it possible to record changes in the size of cities.

From the late 1850s, plates were being engraved which seem never to have been printed from directly – they were used solely as a source of lithographic transfers. A much finer line and neater lettering could easily be achieved, and lithography was less expensive than copperplate engraved maps.

Moreover, colour printing came to play a more prominent role in mapping. Multicolour printing from more than one plate had become possible with the advent of engraving, but it was never common. In the 19th century, the colouring of maps ceased to be a manual process and was transformed by colour

Metropolis

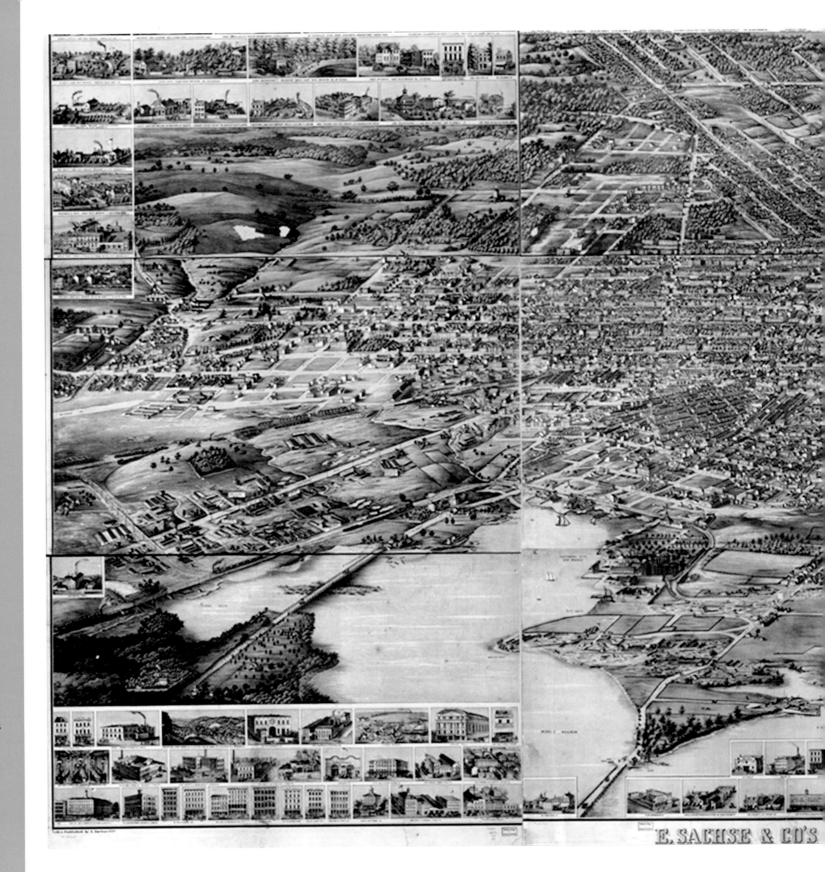

BIRD'S-EYE VIEW OF THE CITY OF BALTIMORE, BY EDWARD SACHSE, 1869 This monumental work (some five feet by eleven) took German immigrant lithographer Sachse and his team of four artists more than three years to complete. It was financed by public subscription and by selling advertising space in the map border to more than a hundred local businesses. Unfortunately, while it is a triumph of cartography it was not a financial success (less than a dozen copies of the map are known to exist today).

The boundary of the bustling commercial–industrial city of Baltimore at the time was from Northern Avenue (modern-day North Avenue) in the north down to the harbour on the tidal part of the Patapsco River in the south, and from Canton in the east to Gwynns Run in the west.

Sachs is believed to have recorded every building, bridge, church, business, park and square in the city, with accurate detail and scale. A prominent exception is Baltimore City Hall, near the centre of the map, which was being built when Sachse's project was in progress. He depicts City Hall's frontage facing south, but when it was completed in 1875 the entrance faced east. **RIGHT**

printing, which increased the amount of information that could be conveyed. This made the use of maps as an explanatory device easier, opening the way for the plentiful production of thematic maps. In addition to – or in place of – the cartographers' traditional interest in topography and elements relating to navigation, other interesting information about a place began to be included, such as health and population density.

Multicolour printing presented more challenges for both mapmakers and users. There was now more

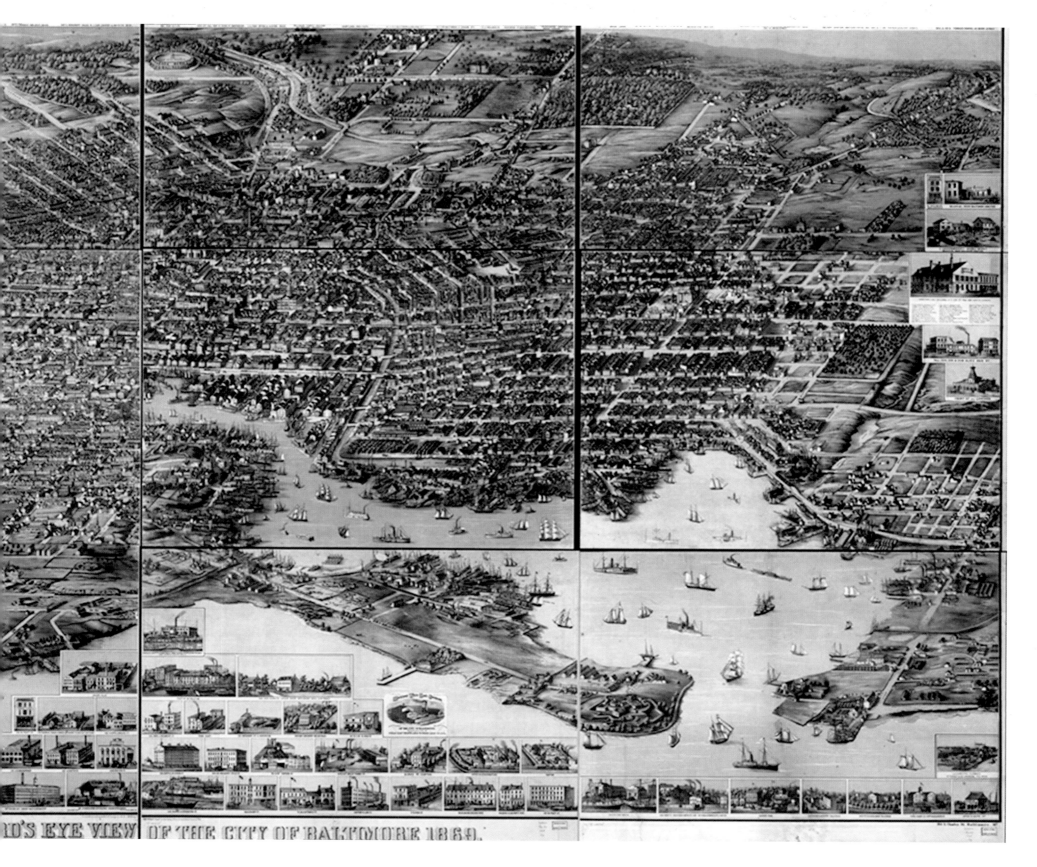

D'S EYE VIEW OF THE CITY OF BALTIMORE 1869.

information to assimilate in the average map, not least through a process of separating out the components, then integrating them in a comprehensible form.

The ability to serve a mass market also helped to create several specific markets for such maps. Universal systems of education led to a schools' market for maps. There was also a considerable growth of interest among the general public: populations were growing, along with leisure interests and discretionary spending, and education helped greatly to increase literacy rates.

POCKET MAP OF THE CITY OF HOUSTON, ISSUED
BY WM. W. THOMAS, 1890 Houston, Texas, is
today the fourth largest city in the USA,
but in the late 19th century the area was
still being developed. William Thomas, a
leading local real estate agent, distributed
this pocket map throughout the USA in
his attempt to market and build Houston
into the hub of Texas.

The area to the northwest of the
centre, bounded by rivers, is now a
cluster of historic districts, dominated by
Houston Heights, which in 1891 became
the earliest planned community in Texas.
It was near Houston but not officially
part of it until annexed by the city in 1919,
and now it forms a district filled with
architectural character. The lots were of
varying sizes to encourage a social mix
and many of the residential plots were
oriented so that buildings faced east or
west to counter the city's hot and humid
summer climate. Retail, commercial and
industrial areas were all planned in from
the outset. RIGHT

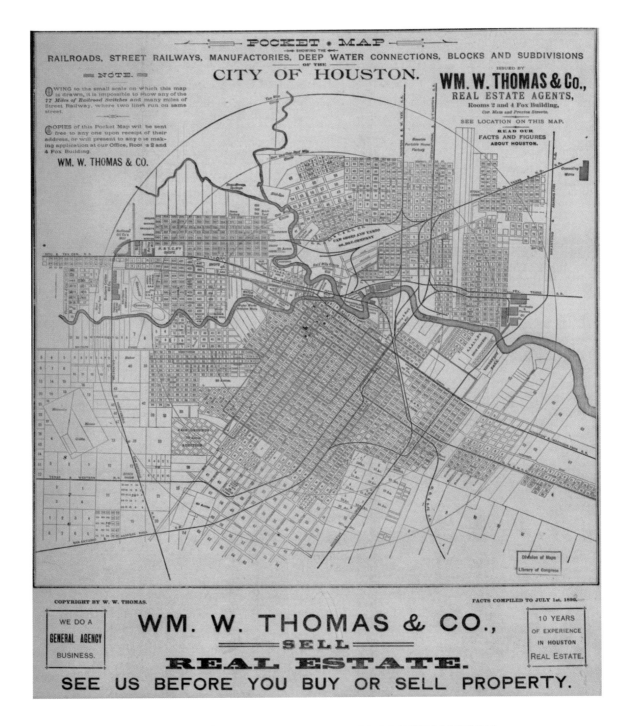

Throughout the era the production of maps
remained an essentially urban activity. Although there
were cities of relatively minor global rank where maps
were produced – for example, Edinburgh, the base of
the W. & A.K. Johnston firm and of Bartholomew and
Son; and Gotha, the base of Justus Perthes – on the
whole, it was either capital cities that were crucial, or
leading cities, such as New York and Philadelphia. The
key global centres of cartographic activity were
London, Paris and New York; the leading publishers
included the Arrowmith family concern and George
Philip and Son in London, and Cotton in New York.

AN AGE OF INFORMATION

The development of thematic mapping was an
especially important one, throwing much needed light
on life in cities, where the contrasts of wealth and
poverty were etched most strongly. Charles Booth's
'Maps Descriptive of London Poverty' (1889) were the
most distinctive product of his *Inquiry into Life and
Labour in London*, representing an attempt to combine
empirical data with time, space and social class.

Booth (1840–1916), wealthy from his Liverpool-
based Booth Steamship Line, was a noted statistician
and president (1892–1894) of the Royal Statistical

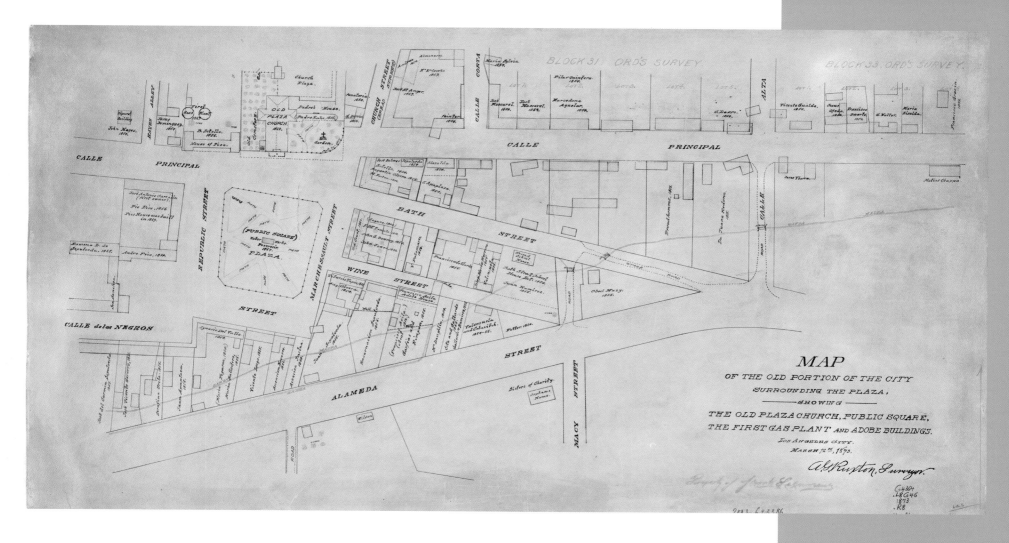

Within the map:

MAP
OF THE OLD PORTION OF THE CITY
SURROUNDING THE PLAZA,
— SHOWING —
THE OLD PLAZA CHURCH, PUBLIC SQUARE,
THE FIRST GAS PLANT AND ADOBE BUILDINGS.
LOS ANGELES CITY.
MARCH 12th, 1873.
A. G. Ruxton, Surveyor.

Society. He doubted the 1886 report of the Social Democratic Foundation that a quarter of the working population were living in poverty, only to find instead that the percentage was higher than that (30.7).

Attitudes then were, of course, different to those which prevail today and Booth's 1889 map ascribed a low moral status to the poorest residents, who were described as 'Lowest class, vicious, semi-criminal', and indicated using black. These were differentiated from two other classes of poor, indicated using shades of blue. The maps showed a clear pattern of spatial division: whereas the sheet for West London's Belgravia contained swathes of bright yellow coding, denoting the wealthy, there were none such for East London's Spitalfields, Wapping and Whitechapel.

In addition to recording and representing working-class poverty, there was also concern about the effects of housing quality, overcrowding, and so on. In Copenhagen, an appreciation of the accommodation issue resulted in the construction of inexpensive housing association blocks for workers in the late 19th century. In Berlin, a major industrial and transport centre – and, from 1871, the capital of a German empire – the population tripled to 870,000 between 1850 and 1870. Large numbers of tenements, or *mietskasernen* ('rental barracks') were built without indoor plumbing. The mid-century Schmidt Plan for the city's managed expansion was given tangible form by James Hobrecht a decade later (page 132).

A WORLD OF CHANGING CITIES

Although mapping was predominantly focused on Western cities, non-Western cities were of increasing importance as well as of interest, even if only for their exotic difference. The ambiguous nature of cities was shown in their role in slavery (one of the century's major moral and social issues), organising the slave trade and providing centres of slave labour, but also being places of freedom to which slaves could flee. Thus, in 1811 Timbuktu in West Africa was described by Robert Adams, an enslaved American sailor, in terms of slave raiding, and in 1873 Russian conquerors

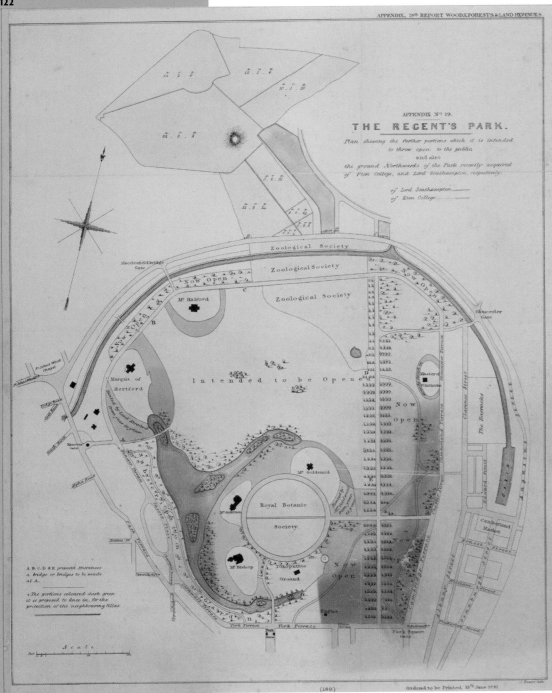

APPENDIX, 18th REPORT WOODS,FORESTS,& LAND REVENUES.

APPENDIX Nº 19.

THE REGENT'S PARK.

Plan shewing the further portions which it is intended
to throw open to the public
and also
the ground Northwards of the Park recently acquired
of Eton College, and Lord Southampton, respectively.

of Lord Southampton
of Eton College

PLANS FOR LONDON IMPROVEMENTS, BY JOHN NASH, 1812 London has had few grand plans in its history, and the best known of those, after the Great Fire of 1666, was not fully enacted. However, by the early 19th century, with London increasingly of global importance, there was a renewed desire to impose order on the organic jumble of the city's streets at the same time as meeting a yearning for healthy, landscaped, open-air spaces. ▶

liberated 30,000 slaves in Khiva in Central Asia. In 1888 Brazil was the last country in the Western world to abolish slavery, by which time more Africans had been taken there than to all of North America and the Caribbean combined. Moreover, slavery in Brazil had always had an urban character: as early as 1550, slaves made up just under 40 per cent of the population in Rio de Janeiro. At the time of abolition, the proportion of slaves in Rio was slightly higher than it had been three centuries earlier – the greatest urban concentration of slaves since ancient Rome. In the

neighbourhoods of Saúde and Gamboa in the Port area of downtown Rio about four million slaves stepped ashore onto the city's wharves, to be traded at Valongo Wharf or in the street slave markets. Ten times as many slaves came to Rio than entered all of the USA. The first of Rio's famous *favela* districts, where samba music originated, was developed by former slaves in the late 19th century in the port area's Morro da Providencia, which became known as 'Little Africa'.

The changing global order had major consequences for cities. In 1842, after British forces advanced to near Nanjing, the Chinese conceded the opening up to trade of what were called Treaty Ports, including Shanghai, Amoy, Fuzhou and Canton (Guangzhou). In 1853, Tokyo was menaced by a US fleet, while in 1860 Beijing was captured by an Anglo-French army during the second Opium War. In 1900, the Boxer Rising led to the foreign legations in Beijing being besieged until their eventual relief by an army of Western and Japanese forces. In 1882, Alexandria and Cairo in Egypt were conquered by the British; in 1898 Khartoum in the Sudan followed. Even Lhasa in Tibet was occupied in 1904. The profits of empire had led to the building of grandiloquent edifices in many European cities, such as Brussels, but industrial wealth was becoming more and more important.

The major cities of the West, which had become increasingly prominent in the world order, menaced each other but were not under threat from non-Western armies. The symbolic power of cities was shown by them hosting great displays of technology and power. London's Great Exhibition of 1851 was followed in New York in 1853 by the second World Fair, with its own Crystal Palace built on the site of what is now Bryant Park. The Egyptian Revival Croton reservoir provided the monumental iron-and-glass structure with a dramatic backdrop.

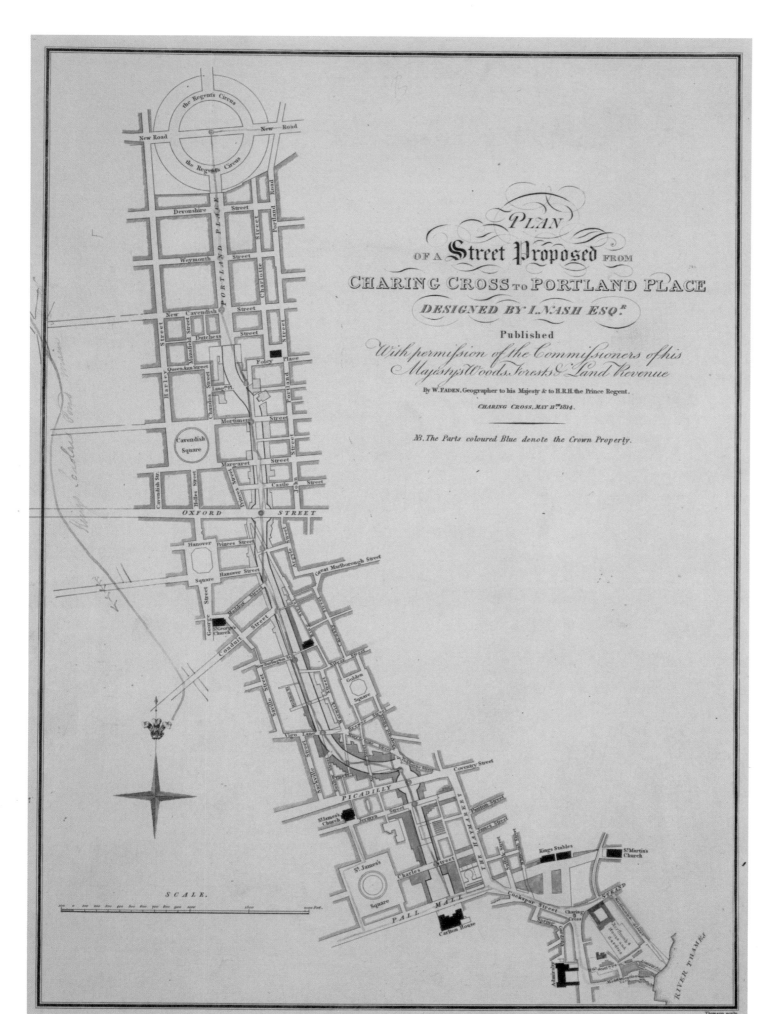

A natural division had arisen between the city's crowded, gritty, industrial East and its affluent West, with fears in some quarters that the East's poor might swamp the West's wealthy. Against this backdrop, the town planner John Nash, under the patronage of the Prince Regent, was invited to devise a series of 'improvements' to enhance the city. His centrepiece was a grand north–south avenue intended to link an exclusive residential park development, including a summer palace for the Prince Regent, to the prince's residence in the heart of the city at Carlton House.

Most of the scheme was never built but several key things did arise from the work: Regent's Park was created, which by the 1820s was home to the Royal Botanic Society and a menagerie for the London Zoological Society (which became London Zoo), with attractive terraces of villas; All Souls church on Langham Place was built as the hinge to link to the existing Portland Place and onward into the elegent, new sweep of Regent Street; and, crucially, a number of slums in the Oxford Street and Soho areas were swept away and the East End and West End became spatially fixed. **LEFT AND OPPOSITE**

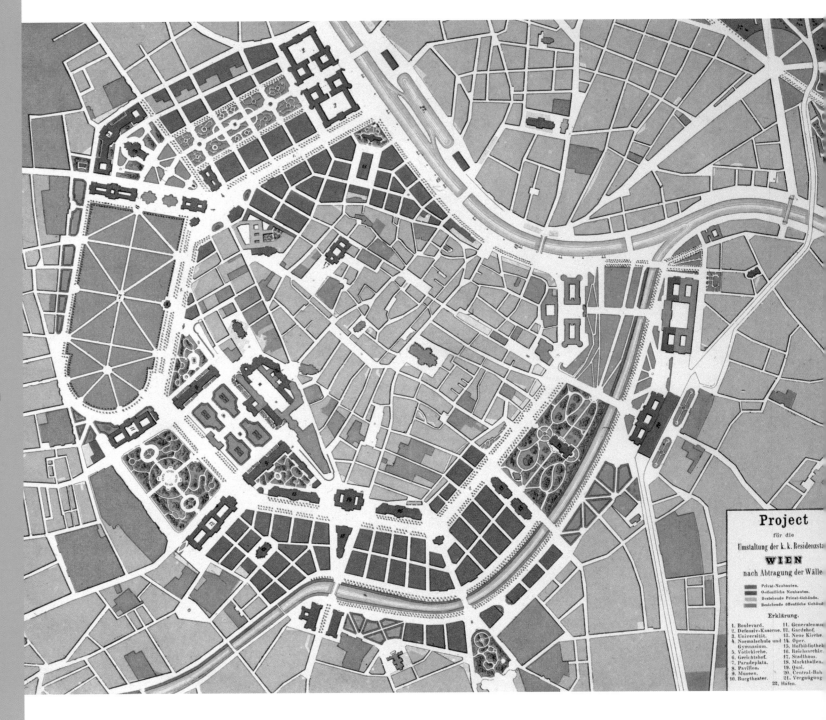

Project
für die
Umstaltung der k. k. Residenzsta
WIEN
nach Abtragung der Wälle

Privat-Neubauten.
Oeffentliche Neubauten.
Bestehende Privat-Gebäude.
Bestehende öffentliche Gebäu

Erklärung:

1. Boulevard.   11. Generalcomm
2. Defensiv-Kaserne.   12. Gardehof.
3. Universität.   13. Neue Kirche.
4. Normalschule und   14. Oper.
   Gymnasium.   15. Hofbibliothek
5. Votivkirche.   16. Reichsarchiv.
6. Gerichtshof.   17. Stadthaus.
7. Paradeplatz.   18. Markthallen.
8. Pavillon.   19. Quai.
9. Museen.   20. Central-Bah
10. Burgtheater.   21. Vergnügung
                        22. Häfen.

THE VIENNA RINGSTRASSE PLAN, BY MARTION KINK, 1859 Many urban centres in Europe underwent transformations during the 19th century, seeking forms of development that were more modern than medieval in character. The popular view was that the age of towns as fortresses had passed, and in 1857 Emperor Franz Joseph decided to raze Vienna's defensive walls, which made land available for planning purposes. The resulting Ringstrasse (1858–1865), overseen by an architectural commission which included Ludwig von Förster, was dominated by a showpiece promenading boulevard that provided a focal point for new public buildings and private homes. **RIGHT**

COPENHANGEN, LITHOGRAPH BY EMIL BAERENTZEN, 1853 Until the middle of the 19th century Copenhagen was essentially a defensive citadel with moats, ramparts and gates, as this period map by artist and lithographer Emil Baerentzen shows. A British bombardment in 1807, a bloodless revolution, a population that was confined in overcrowded conditions that created public health issues (in 1853 cholera killed nearly 5,000 people), all produced pressure to modernise and expand the city. Under town planner Ferdinand Meidahl, moats were changed into parks and, most significantly, the city sought to solve its sanitation problem with a programme of residential accommodation in new suburbs. **OPPOSITE**

It was under the impact of Western rule or influence that non-Western cities acquired many of the features of their counterparts, such as railway stations, boulevards, telegraph buildings and major hotels. In 1888, Sirkeci station was built in Constantinople as the eastern terminus for the Oriental Express, representing a different value of utility to the mosques and palaces that otherwise dominated the cityscape.

Meanwhile, the Western imperial powers were both remodelling existing cities to reflect their own priorities or were developing new cities that would one day play important roles in a new world coming into being. In 1819 the British established Singapore as a deep-water harbour, and by 1860 it had a population of 80,000. Kuala Lumpur was transformed from a shantytown for tin-miners into a city of 40,000 by 1900, with grand, impressive buildings on the pattern of British India, where, in the case of Calcutta, the second largest city of the British empire, there were a series of official buildings in the governmental quarter, such as the High Court (1872). Seized by Britain in 1841, Hong Kong was developed into a leading port

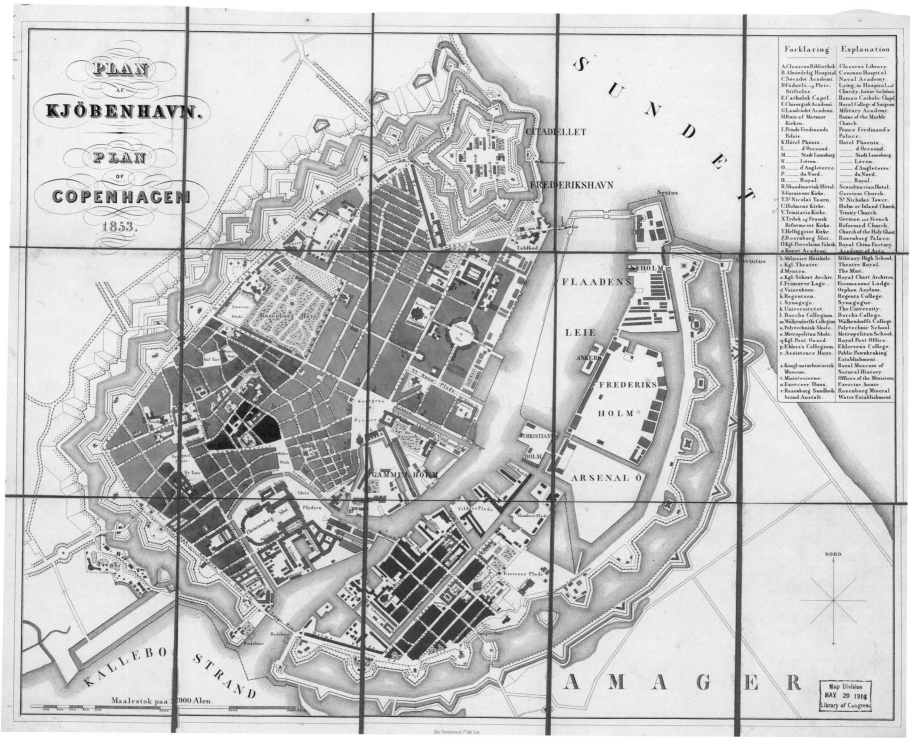

Plan af

KJÖBENHAVN.

PLAN

OF

COPENHAGEN

1853.

Map Division
MAY 20 1916
Library of Congress.

Forklaring	Explanation
A.Classens Bibliothek.	Classens Library.
B.Almindelig Hospital.	Common Hospital.
C.Søcadet Academi.	Naval Academy.
D.Fødsels- og Pleie- Stiftelse	Lying-in Hospital and Charity-house for Infants.
E.Catholsk Capel.	Roman Catholic Chapel.
F.Chirurgisk Academi.	Royal College of Surgeons.
G.Landcadet Academi.	Military Academy.
H.Ruin af Marmor Kirken.	Ruins of the Marble Church.
I.Prinds Ferdinands Palais.	Prince Ferdinands Palace.
K.Hôtel Phönix.	Hotel Phoenix.
L. — d'Oresund.	— d'Oresund.
M. — Stadt Lauenburg.	— Stadt Lauenburg.
N. — Löven.	— Löven.
O. — d'Angleterre.	— d'Angleterre.
P. — du Nord.	— du Nord.
Q. — Royal.	— Royal.
R.Skandinavisk Hôtel.	Scandinavian Hotel.
S.Garnisons Kirke.	Garrison Church.
T.St Nicolai Taarn.	St Nicholas' Tower.
U.Holmens Kirke.	Holm or Island Church.
V.Trinitatis Kirke.	Trinity Church.
X.Tydsk og Fransk Reformeert Kirke.	German and French Reformed Church.
Y.Helliggeist Kirke.	Church of the Holy Ghost
Z.Rosenborg Slot.	Rosenborg Palace.
O.Kgl. Porcelains Fabrik	Royal China Factory.
a.Kunst Academi.	Academy of Arts.
b.Militaire Høiskole.	Military High School.
c.Kgl. Theater.	Theatre Royal.
d.Mynten.	The Mint.
e.Kgl. Søkort Archiv.	Royal Chart Archives.
f.Frimurer Loge.	Freemasons' Lodge.
g.Vaisenhuus.	Orphan Asylum.
h.Regentsen.	Regents College.
i.Synagoge.	Synagogue.
k.Universitet.	The University.
l.Borchs Collegium.	Borchs College.
m.Walkendorffs Collegium	Walkendorffs College.
n.Polytekniske Skole.	Polytechnic School.
o.Metropolitan Skole.	Metropolitan School.
q.Kgl. Post Gaard.	Royal Post Office.
p.Ehlers's Collegium.	Ehlersens College.
r.Assistence Huus.	Public Pawnbroking Establishment
s.Kongl. naturhistorisk Museum.	Royal Museum of Natural History.
t.Ministerierne.	Offices of the Ministers.
u.Exerceer Huus.	Exercise-house.
v.Rosenborg Sundheds- brønd Anstalt.	Rosenborg Mineral Water Establishment.

Maalestok paa 2000 Alen.

and city. Government House was finished in 1855 and the Hong Kong and Shanghai Bank, established in 1864, played a key role in financing trade.

Steam transport, in the shape of steamships and railways, was also important to the spread of Western settlement and the foundation of new cities, many of which developed as ports and railway centres. For example, Vancouver began in 1884 when the Canadian Pacific Railway decided to make it the terminus of its transcontinental railway. Steamships and railways also took prospectors on the gold rushes, such as the one to California in 1849, which resulted in levels of settlement that helped propel the growth of cities such as San Francisco, creating opportunities that in turn drew immigrants from Asia. The world was becoming interlinked through its growing cities.

Metropolis

MAP OF LIVERPOOL, BY JOHN TALLIS IN 1851

The map of Liverpool is enlivened by a
full-width view from across the River
Mersey in Birkenhead, showing the city's
mid-Victorian-era skyline. Decades of
commercial growth had resulted in the
creation, under dock surveyor Jesse
Hartley, of the world's first fully enclosed
wet dock system that added ten miles of
quay space and 140 acres of docks to the
system, including the Albert Dock
(1843–47). By the 1840s Samuel Cunard
was operating his transatlantic service.

The wealth from this trade resulted in
new residential park suburbs, such as
Princes Park (1842, visible to the right side
of the map) and many fine public
buildings. Part of the appeal of Tallis's
map are the vignettes decorating the
border: steamboats, sailing ships and,
clockwise from top right, St George's Hall,
the Sailors' Home and the Custom House.
The city's population had quadrupled to
more than 300,000 and along with great
wealth came unparalleled squalor, which
led to Liverpool becoming one of the
first places anywhere to embark on a
programme of municipal house building.

OPPOSITE

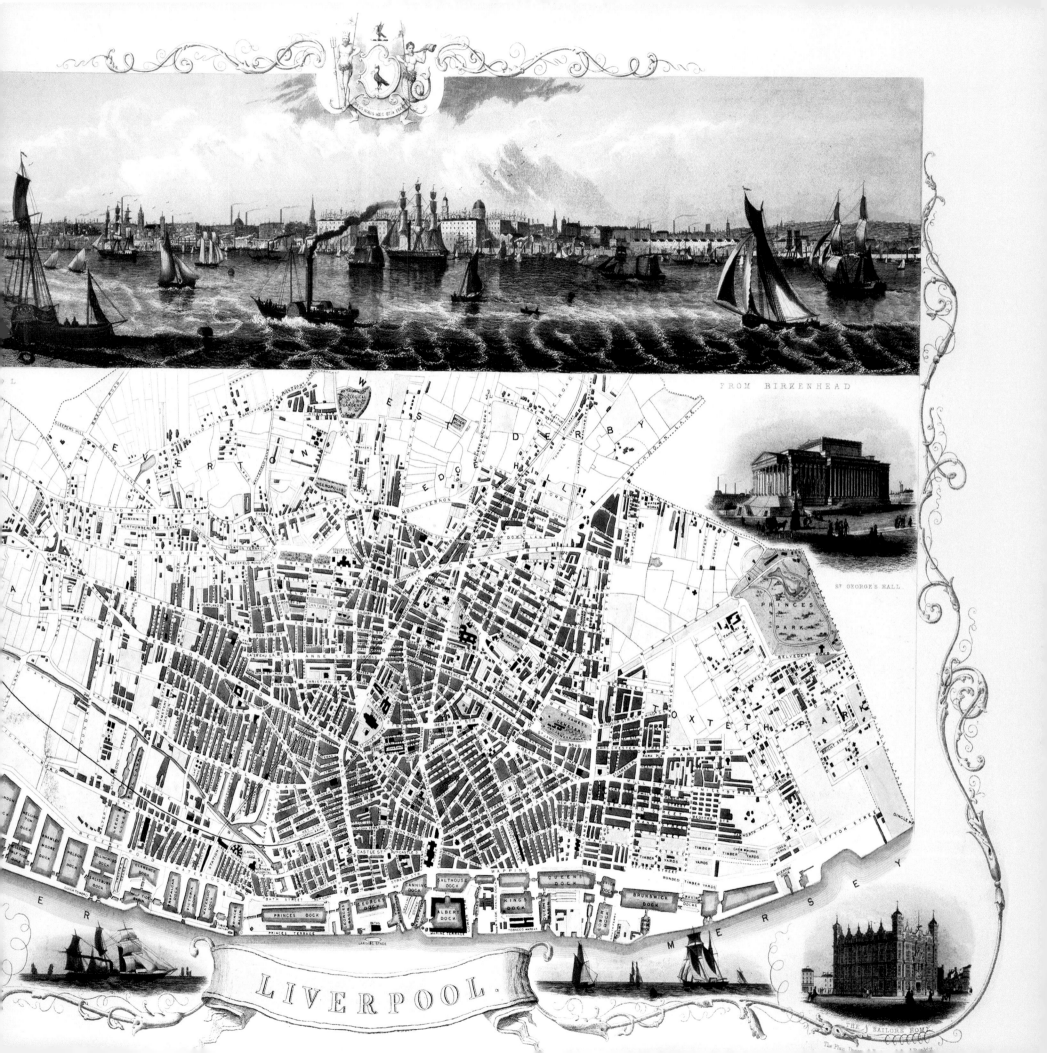

FROM BIRKENHEAD

ST GEORGE'S HALL.

LIVERPOOL.

THE SAILORS HOME

BIRD'S-EYE VIEW OF THE CHICAGO PACKING HOUSES & UNION STOCK YARDS, BY CHARLES RASCHER, 1890 This fascinating perspective print shows the streets, buildings and railroads of Chicago's 'meatpacking' district, a notorious area built on swampland to the south of the city in the mid-1860s, where Upton Sinclair later set his novel *The Jungle* (1906). The transport and butchering of livestock for human consumption was a huge industry for a century (until the yards closed in 1971), employing thousands of workers who processed as many as nine million animals per year in the yards by 1890. **RIGHT, TOP**

KANSAS CITY, BY AUGUSTUS KOCH, 1895 This bird's-eye panorama depicts the city's West Bottoms district with its stockyards, railyards and factories, with the Missouri River in the foreground and the Kaw River in the distance. Reid Brothers Packing Company (inset) was one of many meatpacking companies in this part of the city and neighbouring Armourdale which believed the proximity of Kansas City to the grasslands of the Midwest meant that it could wrest the title of chief meat-processing centre from Chicago. However, the low-lying area was subject to flooding and only a decade later many companies had moved. Beyond the railway, cattle can be seen (upper left of centre) tightly packed within their griddled feedlots. **RIGHT, BOTTOM**

CHICAGO BIRD'S-EYE FOR ROCK ISLAND AND PACIFIC RAILWAY, BY POOLE BROTHERS, 1897. From the 1860s elevated railways became popular in US cities and this finely detailed view of Chicago from the lakefront depicts its loop railway, with a focus on Rock Island Station. The lakefront area had changed since Palmatory's view in mid-century with the offshore rail trestle having gone since the Grant Park district became more unified and apparent. **RIGHT**

ALL ELEVATED TRAINS IN CHICAGO
STOP AT THE
Chicago Rock Island AND Pacific Railway Station
ONLY ONE ON THE LOOP

ROCK ISLAND ROUTE

ELEVATED STATION AT THE "ROCK ISLAND" VAN BUREN ST. STATION.

CHICAGO, ROCK ISLAND & PACIFIC RAILWAY.

TICKET OFFICES AND STATIONS.

Van Buren St. Station	"	A
22d Street	"	B
Englewood	"	C
City Ticket Office	"	D

OTHER RAILWAY STATIONS.

Central	Station	L— 8
Dearborn St.	"	J—14
Nickel Plate R.R.	"	I—14
Grand Central	"	H—19
Union	"	H—29
C. & N.W. Ry.	"	N—37

HOTELS.

Auditorium	N—13
Auditorium Annex	N—12
Atlantic	J—21
Briggs	M—31
Brevoort	N—37
Burke's	N—26
Clifton	R—20
Commercial	S—30
Grand Pacific	L—22
Great Northern	N—20
Grace	M—20
Gore	K—20
Leland	Q—15
McCoy's	K—20
Merchants'	Q—31
Palmer House	P—20
Sherman	P—29
Tremont	S—28
Victoria	P—14
Wellington	P—16

DRY GOODS AND DEPARTMENT STORES.

Boston Store	Q—24
Carson, Pirie, S. & Co.	R—24
Marshall Field (Wholesale)	J—26
" " (Retail)	T—24
Morgenthau, B. & Co.	P—21
Mandel Bros.	R—23
Rothschild's	N—17
Siegel, Cooper & Co.	M—15
Schlesinger & Mayer	R—22
The Fair	O—21

OFFICE BUILDINGS.

Ashland	P—29
Chamber of Commerce	N—28
Columbus Memorial	R—24
Fisher	M—18
Ft. Dearborn	M—24
Great Northern	N—20
Kedzie	O—26
Masonic Temple	U—20
Marquette	M—22
Monadnock	N—16
Manhattan	L—17
Medinah	J—24
New York Life	M—25
Old Colony	L—17
Owings	N—21
Pontiac	K—16
Rialto	L—22
Rookery	L—23
Stock Exchange	M—25
Title & Trust	P—26
Trude	U—24
Unity	K—21
Woman's Temple	L—25
Y. M. C. A.	M—26

PUBLIC BUILDINGS.

CITY.

Art Institute	S—16
Battery "D"	U—17
Board of Trade	K—26
City Hall	O—28
Open Board	K—21
Public Library	X—22

COUNTY.

Court House	N—29

FEDERAL.

Post-Office, new	M—21
" temporary	W—20
U.S.Bonded Warehouse	I—19

THEATERS AND AMUSEMENTS.

Battle of Gettysburg	L—11
Columbia	N—23
Chicago Opera House	O—27
Great Northern	N—20
Grand Opera House	P—28
Gaiety	Q—27
Hooley's	O—30
Hopkins'	L—14
McVicker's	Q—23
Schiller	Q—28
Sam T. Jack's	Q—23

The Red Lines (light) represent the Cable and Electric Street Car System as connections with the Elevated System in the City.

Copyrighted 1897 by Poole Bros. Chicago.

STEAMER LANDING FOR ALL LAKE POINTS

W. H. TRUESDALE, Vice-President and General Manager.

JOHN SEBASTIAN, General Passenger and Ticket Agent.

MAPPING SOCIETY: a scientific era

By the middle of the 19th century there was increasing concern about living and working conditions within major cities. Social reformers and campaigners argued that extreme poverty, overcrowding and insanitary conditions, which could cause epidemics that affected rich and poor alike, should be eradicated in civilised society. The scientific collection of data and the development of thematic mapping to illustrate a range of issues contributed to a revolution in political, medical and social attitudes.

AN EPIDEMIC IN GLASGOW, BY ROBERT PERRY, 1844

Robert Perry was a physician at the city's Royal Infirmary. His book, *Facts and Observations on the Sanitary State of Glasgow* (1844), used statistical tables and mapping to support the argument that poverty, disease and crime were interconnected. This Glasgow overview map of an influenza outbreak numbers the districts and uses a darker shade to support the evidence that the epidemic was worst in the most densely inhabited areas. Interesting incidental detail includes the livestock markets near Gallowgate, above the Green. **RIGHT**

SANITARY MAP OF THE TOWN OF LEEDS, BY DR ROBERT BAKER, 1842 Edwin Chadwick reproduced this map in his *Report on the Sanitary Conditions of the Labouring Population of Great Britain.* Baker plotted clusters of cholera outbreaks and contagious diseases in 1832 and argued for public cleanliness. Commissioned in 1839 by the government to investigate the lives of the country's poor, Chadwick found that individual immorality or idleness was less to blame than modern urban life, a revelation that prompted programmes of public health reform and schemes for urban improvement. **RIGHT**

Hotbeds of Innovation: 1800–1900

PART OF CHARLES BOOTH'S 'MAPS DESCRIPTIVE OF LONDON POVERTY', 1889 These maps are the most distinctive product of Booth's *Inquiry into Life and Labour in London* (1886–1903). This section shows Spitalfields, Wapping and Whitechapel. Although his text was based on an eight-tier system of social class, the map legend consisted of seven colour codings: 'BLACK: Lowest class. Vicious, semi-criminal. DARK BLUE: Very poor, casual. Chronic want. LIGHT BLUE: Poor. 18s. to 21s. a week for a moderate family. PURPLE: Mixed. Some comfortable others poor. PINK: Fairly comfortable. Good ordinary earnings. RED: Middle class. Well-to-do. YELLOW: Upper-middle and Upper classes. Wealthy.'

A combination of colours – as dark blue or black – indicates that the street contains a fair proportion of each of the classes represented by the respective colours. The lowest class, the pockets of black, he described as: '...some occasional labourers, street sellers, loafers, criminals and semi-criminals. Their life is the life of savages, with vicissitudes of extreme hardship and their only luxury is drink.' Many of the areas of greatest poverty were to be found in housing known as 'courts', behind the street frontages. In the social hierarchy beneath the 'servant-keeping class' (yellow) were the lower middle class, signified by red: 'Shopkeepers and small employers, clerks and subordinate professional men. A hardworking sober, energetic class.' These residents lived along the main streets. **LEFT**

DEVELOPMENT PLAN FOR THE ENVIRONS OF BERLIN, BY JAMES HOBRECHT, 1862 Unlike the contemporary reconfiguration of Paris by Haussmann, the plan that James Hobrecht was commissioned to produce for Berlin did not involve clearing and redeveloping the historic centre. Instead, Hobrecht's was an urban vision that provided guidelines for the future growth and outward expansion of the great city.

He was not the first to notice Berlin's housing problem, poorly developed industrial zones and inadequate transport links, but he was the first to produce a coherent and flexible plan to tackle those issues (in so doing he amalgamated some of the earlier 1850s' Schmidt Plan).

Hobrecht studied the solutions used in Paris, London and Vienna before he proposed dividing the city into 14 manageable administrative districts, where development could be implemented using a street layout devised around standardised housing blocks (*mietskasernen)*, though he did not dictate their aesthetics), orbital distributor roads and main radial roads. The density of the development was to be balanced by interspersing it with regular squares and open spaces. Although some of his ideas were not fully implemented, his plan was the foundation upon which the city expanded in the late 19th century. An impressed Mark Twain praised Berlin as 'the Chicago of Europe'. RIGHT

PLAN von BERLIN und

Übersichtskarte des Bebauungsplanes der

MGEGEND bis **CHARLOTTENBURG**

bungen Berlins.

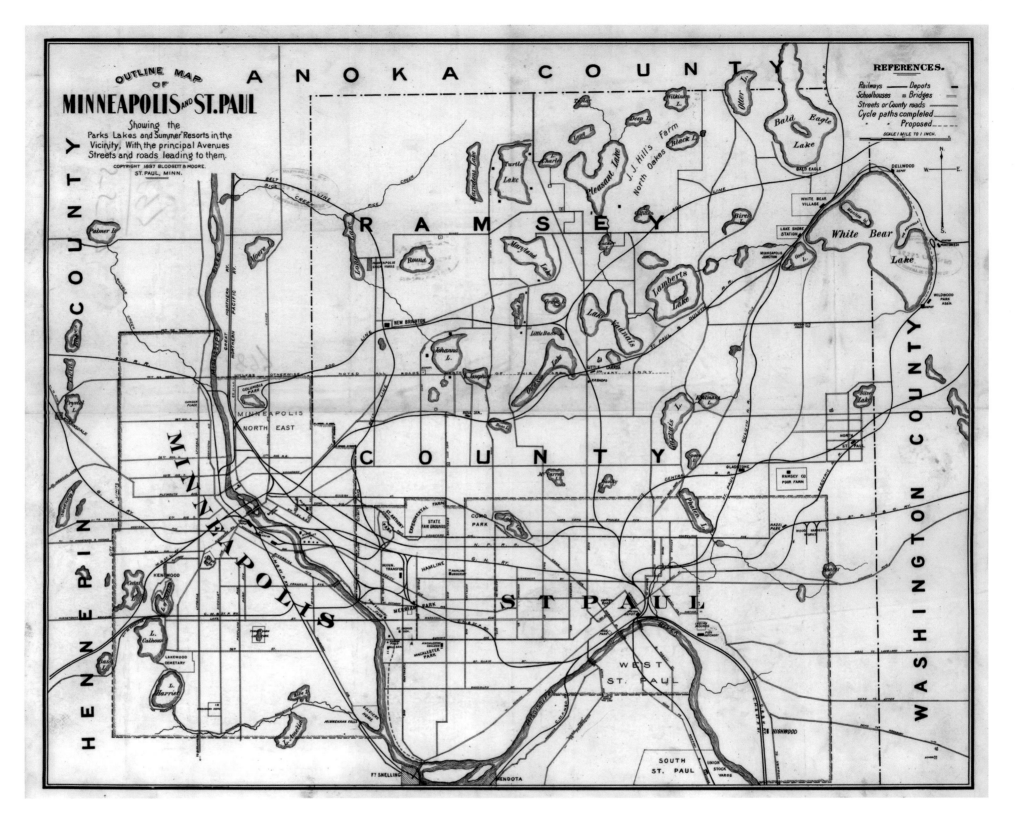

A LEISURE INFORMATION MAP FOR THE 'TWIN CITIES' OF MINNEAPOLIS–ST PAUL, 1897 In the 1890s, following the invention of the 'safety bicycle' in the 1880s, there was a short-lived mania for cycling in the USA, which fuelled a public demand for good roads to be built. In 1895 cyclists in the 'twin cities' of Minneapolis and St Paul funded six miles of dedicated paths and in 1897–1898 the extent of these routes were more than doubled, most of it paid for by individuals through the St Paul Cycle Path Association. Accompanying this rise of interest in outdoor pursuits and leisure time, maps such as this were produced to show cyclists where the best routes lay. **ABOVE**

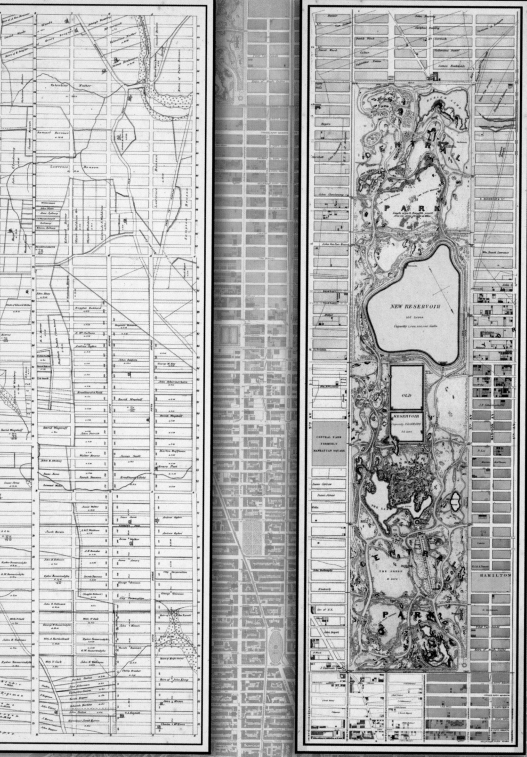

CENTRAL PARK

1815 PLAN

1867 PLAN

NEW YORK CITY

CENTRAL PARK DEVELOPMENT, NEW YORK, BOARD OF COMMISSIONERS, 1867 Before the 1800s common land separated New York's upper Manhattan from the nearby town of Harlem. After the Revolution, the city authorities surveyed and sold off plots, and by 1815 much of the land was owned by families who had settled the area in the 1660s, such as the Waldrons. Roads laid out in that survey were to become Fifth and Sixth Avenues. In the 1850s the idea took hold that New York needed a great park to provide 'urban lungs' for all. After several years of wrangling, in 1854 a central location was chosen. The park reshaped not only the topography of Manhattan but also the city's social landscape; those who had lived on the rocky, swampy site were evicted and what became the affluent Upper East Side district began to take shape. The design competition for the park was won by the naturalistic 'Greensward Plan' of Frederick Law Olmsted and Calvert Vaux, both of whom aspired to a democratic space where the city's classes could mix freely. Begun in 1858, by the mid-1860s the 843-acre (340-hectare) park was largely complete, with its pathways and promenades, forest and lakes, bridges, and architectural structures and rustic features (from rocks to individual trees) – as well as a reservoir that went on to replace (until 1993) the Croton water system (1842–1877), on the site of which now stands the New York Public Library.

LEFT

Metropolis

CHICAGO, BIRD'S-EYE VIEW, BY JAMES PALMATARY AND CHRISTIAN INGER, 1857 A superbly detailed lithograph that shows the lakefront warehouses, grain elevators and lumberyards – a city focused on business. The railway reached Chicago in 1848, and it became the focal point in the USA's rail network, linking the resources of the West to the cities of the East. The train in the foreground is from the Illinois Central Railroad, travelling on a trestle along a breakwater it had to build to reach its lakefront depot, which created a basin that protected many fine mansions. Grant Park has not yet been developed (using landfill from the Great Chicago Fire).

In the centre, south of the river, is the newly built courthouse (1853, later city hall), bordered clockwise north to east by Randolph, Clark, Washington and LaSalle streets. To the north side of the river (right), where the oldest settlers lived, are St James' Episcopal church and Holy Name Catholic cathedral, which burned down in 1871. Inset is a map created to raise money for victims of the fire on 8 October that year. Rather than the area centred on Washington and Clark streets (at the heart of the map's inner circle), the fire began further south, on DeKoven Street between Jefferson and Clinton streets, and it moved north and east, with neither branch of the river impeding its destructive advance. Ultimately the greatest area of fire damage was in the city's northern district. **RIGHT AND INSET**

MAP SHOWING THE BURNT DISTRICT IN CHICAGO!

LAKE MICHIGAN

Published for the benefit of the Relief Fund by
3ᴰ EDITION.   THE R.P. STUDLEY COMPANY, ST. LOUIS.

Metropolis

BOSTON AND WASHINGTON, DC, SANBORN FIRE INSURANCE MAPS, 1867 AND 1888 Insurance maps originated in London in the late 18th century. D.A. Sanborn, a Massachusetts surveyor, founded the Sanborn Map Publishing Company in 1867 after his earliest maps had appeared in a successful atlas called *The Insurance Map of Boston*. It contained 29 maps, surveying property in sections of the city in the detail needed by the fire insurance industry to assess the risk.

Sanborn used a key that combined colours and letters (including blue for stone or concrete buildings, yellow for frame structures and pink for brick; and F for Flat, or D for Dwelling) to identify a building's construction type in detail. 'Specially Hazardous Risks' were green.

This plate of Boston (1867) shows the Albany Street area, where in July 1861 a fire at number 131 (near the carpenter's shop plot), caused by firecrackers on Independence Day, resulted in the loss of up to 20 three-storey buildings extending through to Hudson Street. That day the city had 14 fires causing losses of nearly a million dollars.

The Washington, DC, image is an overview index map from the title page, meaning the colours merely highlight the coverage of individual maps. There is notably limited development in the city to the east and south, and centrally in the area from the Mall and Executive Mansion westwards (although, as Sachse's 1884 view makes clear, there were structures in these areas). **RIGHT AND INSET**

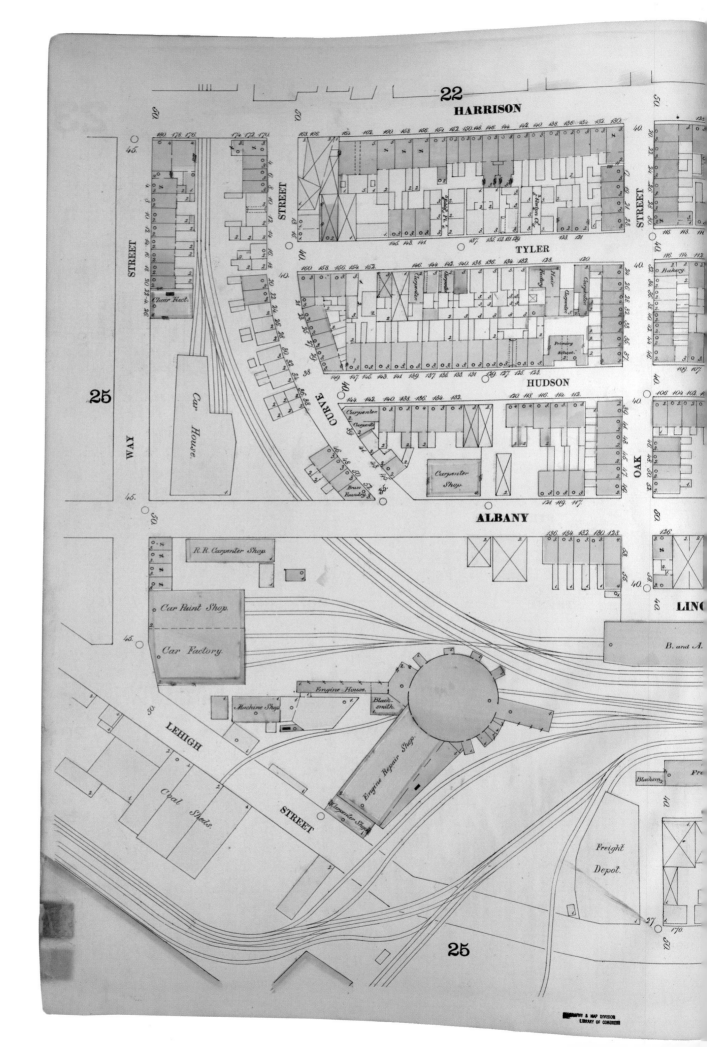

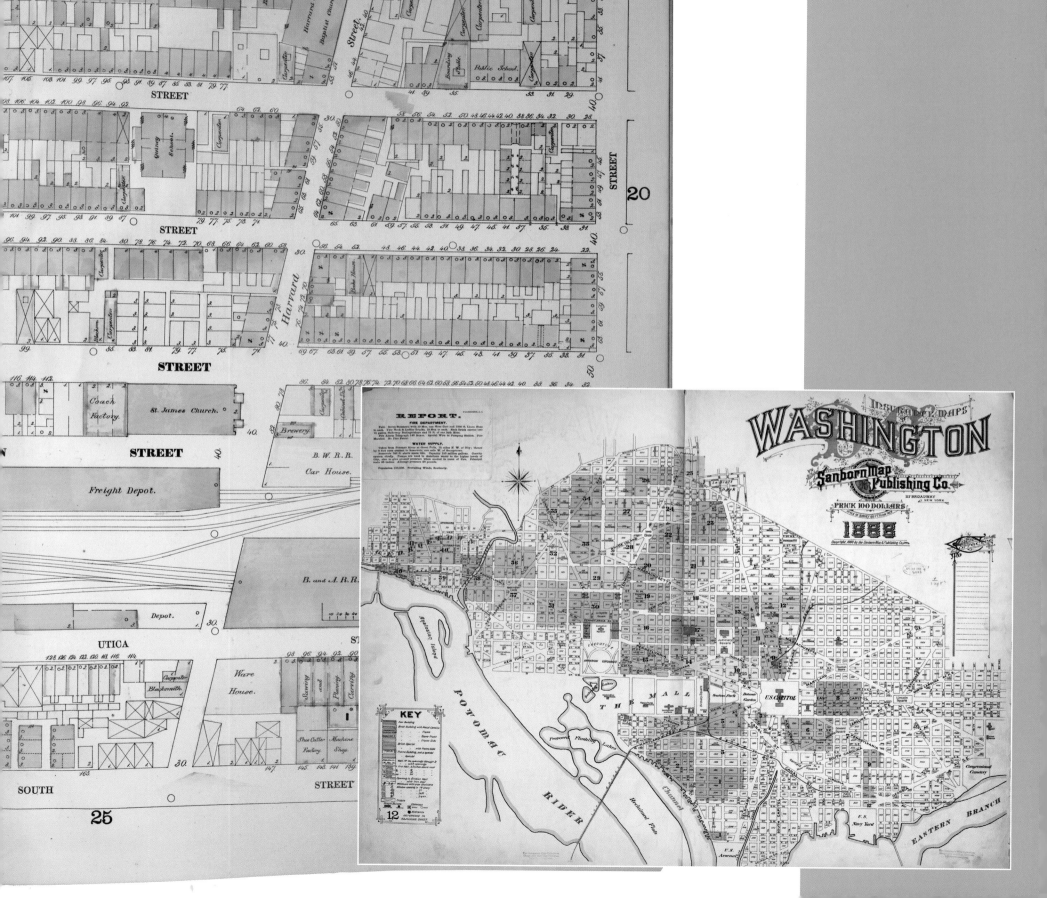

A VIEW OF WALLED NANJING, 1850–1853 This map has north at upper left. During the Taiping Rebellion (1850–1864), in March 1853 Nanjing fell and was renamed Tianjing ('Heavenly Capital'). By far the largest of the city's 13 gates, the complex, castle-style Zhonghua, or Jubao, Gate is shown at bottom right. Qing government forces began a siege of Nanjing from the Yuhuatai encampment. Camp followers are depicted, including foodsellers. Rebels refused to have the Manchu tonsure of shaved head and ponytails, or queues, which explains the camp's barber and the severed heads hanging by their hair nearby. **ABOVE**

HANGZHOU, 1867 Hangzhou is one of China's seven ancient capitals and, with Nanjing, Changchun and Kunming, one of the country's four 'garden cities'. The city's most famous sight – West Lake, with its temples and villas – is instantly recognisable. The then 2,300-year-old Grand Canal split the city into two districts, Jiacheng (the lakeside, western part) and Luocheng (the eastern part). Since 1293 the canal had linked Hangzhou to Beijing in the north, connecting the Yellow and Yangtze river systems. Marco Polo knew the city as Kinsay; awestruck, he referred to a 'City of Wonder', marvelling at its watery setting (it sits on the Qiantang River) and the profusion of bridges, baths and bustling markets. **BELOW**

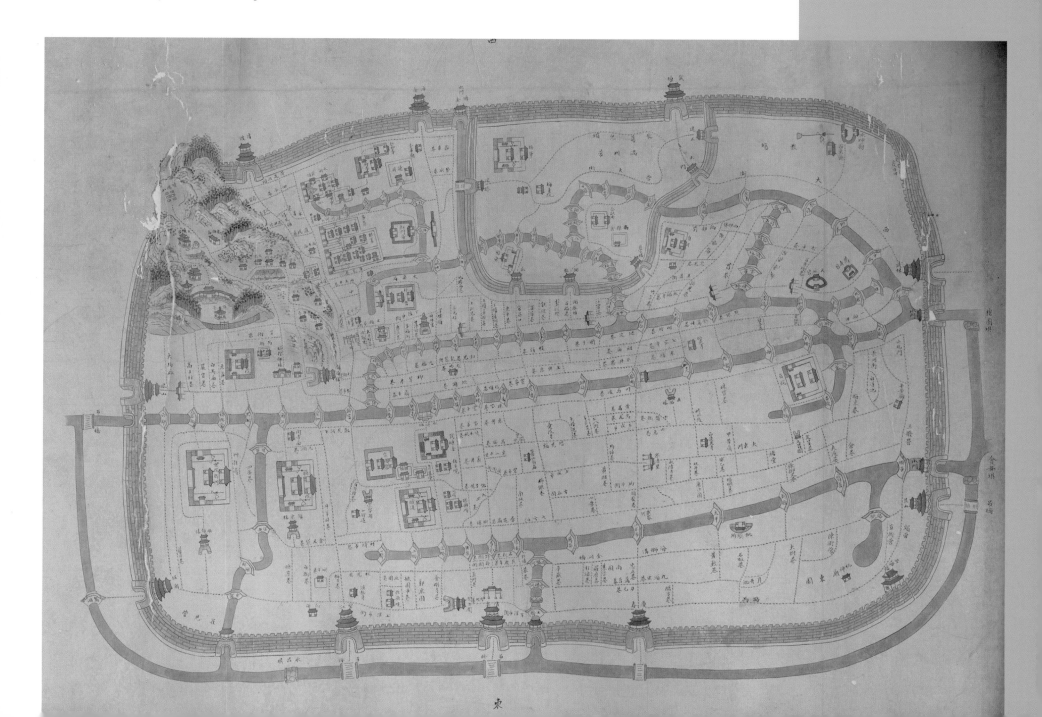

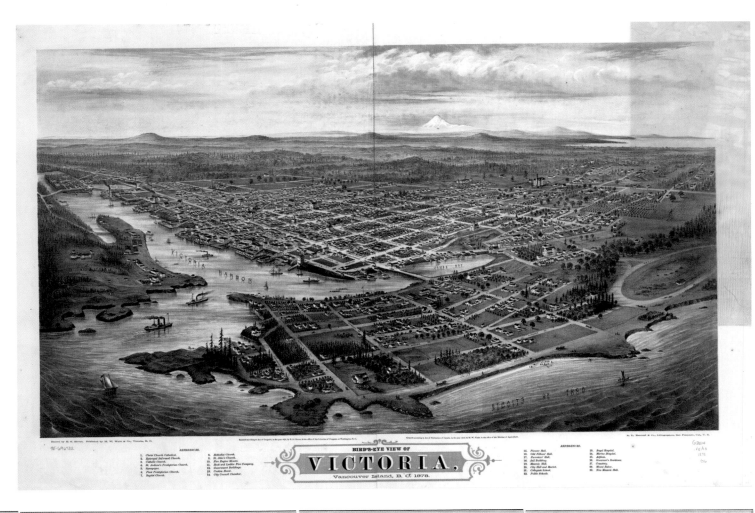

BIRD'S-EYE VIEW OF
VICTORIA,
Vancouver Island, B. C. 1878.

Plan
OF
CALCUTTA.
From Actual Survey
In the years 1847–1849.

Hotbeds of innovation: 1800–1900

MADRAS TOWN AND SUBURBS, SURVEY OF INDIA OFFICE, 1861 Built in 1640 on a leased beach, traders first settled near Fort St George thanks to a tax exemption. The walled township became known as 'Black Town' and from this starting point a city has grown southwards into what many today call Chennai, with what were villages now urban neighbourhoods. **LEFT**

VICTORIA, BRITISH COLUMBIA, BY ELI GLOVER, 1878 Having begun life as a fort at Camosack built by the Hudson's Bay Company, the township was named in honour of the queen in 1846 and grew rapidly from 1858 onwards because of the Fraser River Gold Rush. The harbour and wharf area is to the left of the map, while James Bay is towards the centre. At the bottom a numbered key identifies places of interest, including the Custom House (13) and the City Hall and Market (20). Snow-capped Mount Baker (28) is in the distance. Beacon Hill, now a park, is at bottom right. **OPPOSITE, TOP**

CALCUTTA, BY FREDERICK WALTER SIMS, 1857 Based on data from surveys in 1847–1848, with suburban additions from surveys in 1849, the East India Company map shows roads, significant private and public buildings, temples, public aqueducts, water tanks and drains. Calcutta was the capital of British India until 1911 and had many important buildings. **OPPOSITE, BOTTOM**

Metropolis

PLAN OF THE CITY OF RIO DE JANEIRO, BY E. & H. LAEMMERT, 1867 In the early 1800s many people had left Portugal for Rio and by 1821 the population was 86,000, just under half of them slaves. By 1890 there were no longer slaves and the city had around 420,000 inhabitants. Many of Brazil's institutions could be found in the city, and the port offered the only sizeable dry dock in the region, built in 1861 among the naval facilities on the Ilha das Cobras (Snake Island) in the harbour. From 1859 onwards the existence of a horse-drawn steeetcar service had meant that the city could begin to spread and develop suburbs.

The area is mountainous, including Corcovado (upper left), now the site of the iconic "Cristo Redentor" statue ("Christ the Redeemer"). The map has hachures to show relief, and is sufficiently detailed to have a numbered key that identifies 90 locations and places of interest, from churches and schools to plazas and theatres. **OPPOSITE**

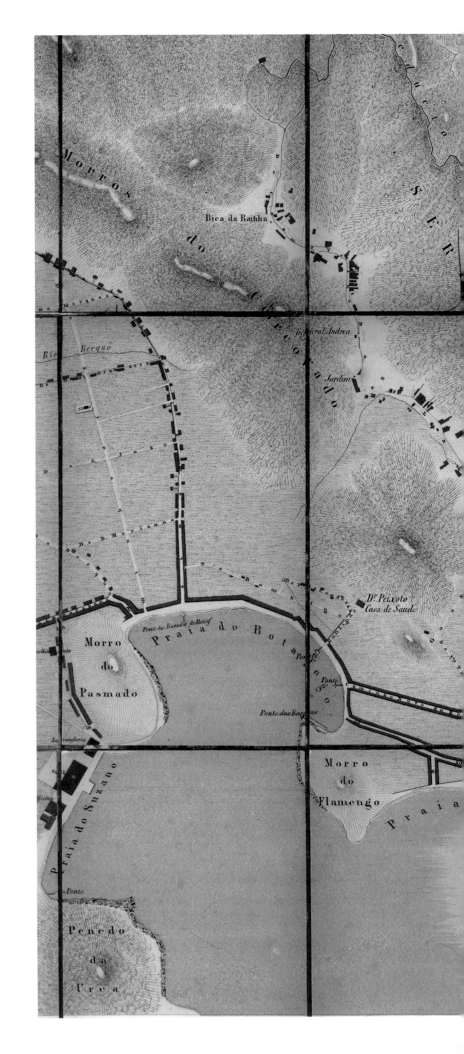

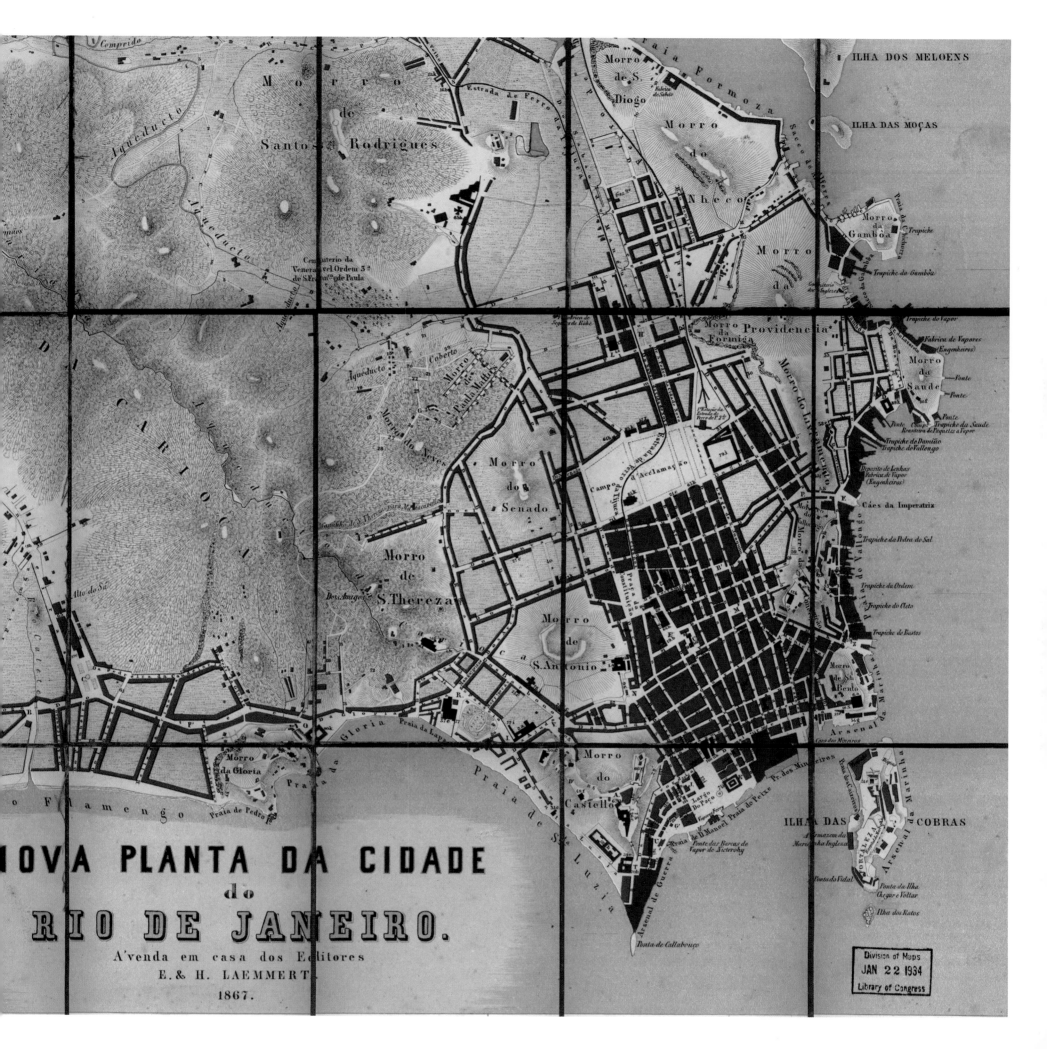

NOVA PLANTA DA CIDADE
do
RIO DE JANEIRO.

A'venda em casa dos Editores

E. & H. LAEMMERT.

1867.

Metropolis

NEW MONUMENTAL PARIS, BY F. DUFOUR, 1878 Now connected by the railways that were expected to bring visitors, this map was intended as a tourist guide to the city's sights. Many of these were new because Paris had been massively altered by a programme of modernisation under Georges Haussman, Prefect of the Seine from 1853 to 1870. Streets were widened, new boulevards and bridges built, and parks laid out, including Parc Montsouris and Parc des Buttes Chaumont (just inside the walls to the southeast and northeast, respectively). The Arc de Triomphe also gained seven new radiating avenues to add to the existing five. BELOW

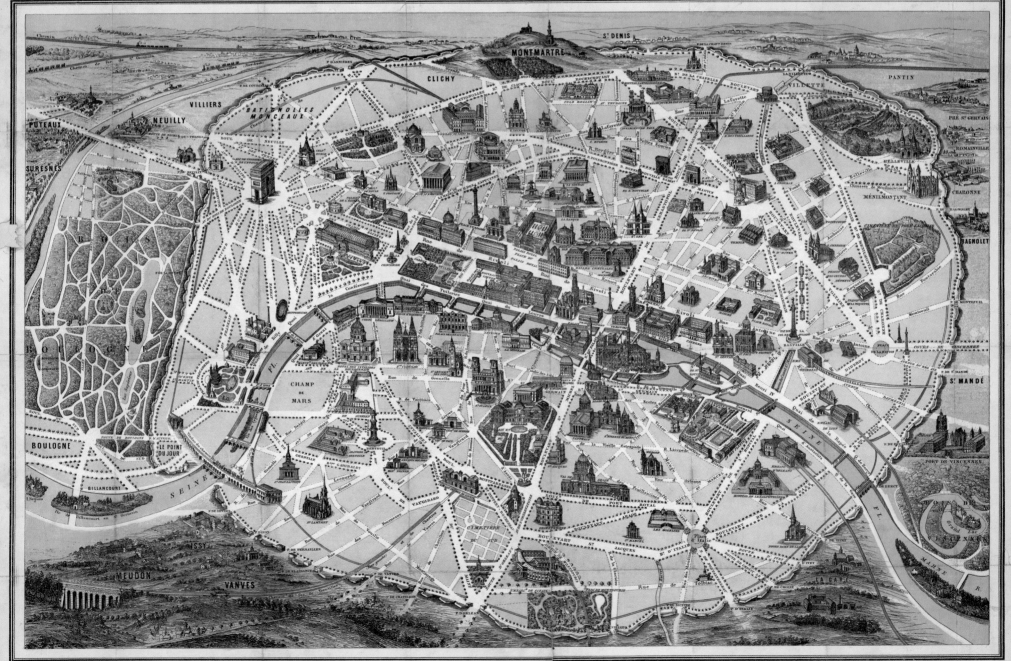

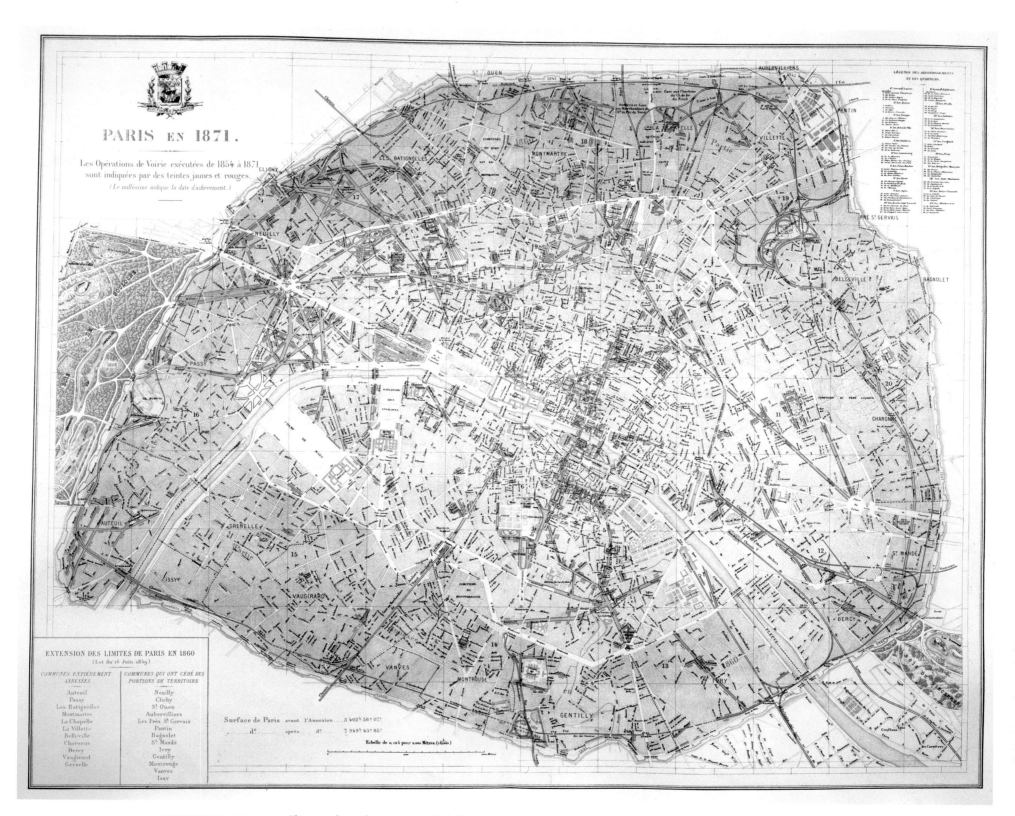

PARIS IN 1871, BY LOUIS WUHRER This map shows the extension (shaded in beige) of the city limits made in June 1859 to incorporate all or parts of 24 suburban communes, creating 20 arrondissements. The loss in 1860 of the wall of the farmers-general (built in 1784–1791), being a toll boundary rather than a defensive wall, was met with indifference. After the annexation some urban areas that had fallen out of favour, such as Notre-Dame-des-Champs (near the Luxembourg Gardens), were gentrified and new ones became popular with the middle class, such as Parc Monceau, where in 1871 many Communards were massacred. **ABOVE**

Metropolis

SHANGHAI, BY THE DIANSHIZHAI STUDIO, 1884

This is a map of foreign 'concessions' in Shanghai – the districts are colour coded, now rather faded, to indicate the nationality, from north to south: orange (US), blue (British, 1843), red (French). Beneath the French area is the Chinese (walled) part of the city, highlighted in yellow. The *Dianshizhai Pictorial* appeared in the last quarter of the 19th century (1884–1898), reflecting the social change and cosmopolitanism affecting the city. Although foreign-owned, the newspaper was written in Chinese for a local readership; extraterritoriality meant the editorial freedom to ignore the views of the Chinese government. **RIGHT**

TIANJIN TREATY PORT, BY FENG QIHUANG, 1899

Oriented with north at upper left, this watercolour map depicts the walled city of Tianjin (or Tientsin), on the edge of the North China Plan. In 1858, after the Second Opium War, Tianjin became an open trading port with concessions to Britain and France, and in 1895–1902 other foreign nations. As a result, the city is shown with factories and railroads near the Hai River and Grand Canal. The core rectangular wall dates from the Ming era, when in 1404 the original settlement became a garrison town. Its location at the northern end of the Grand Canal meant it provided the gateway to Beijing and prospered under the Qing from 1644 onwards. **BELOW OPPOSITE**

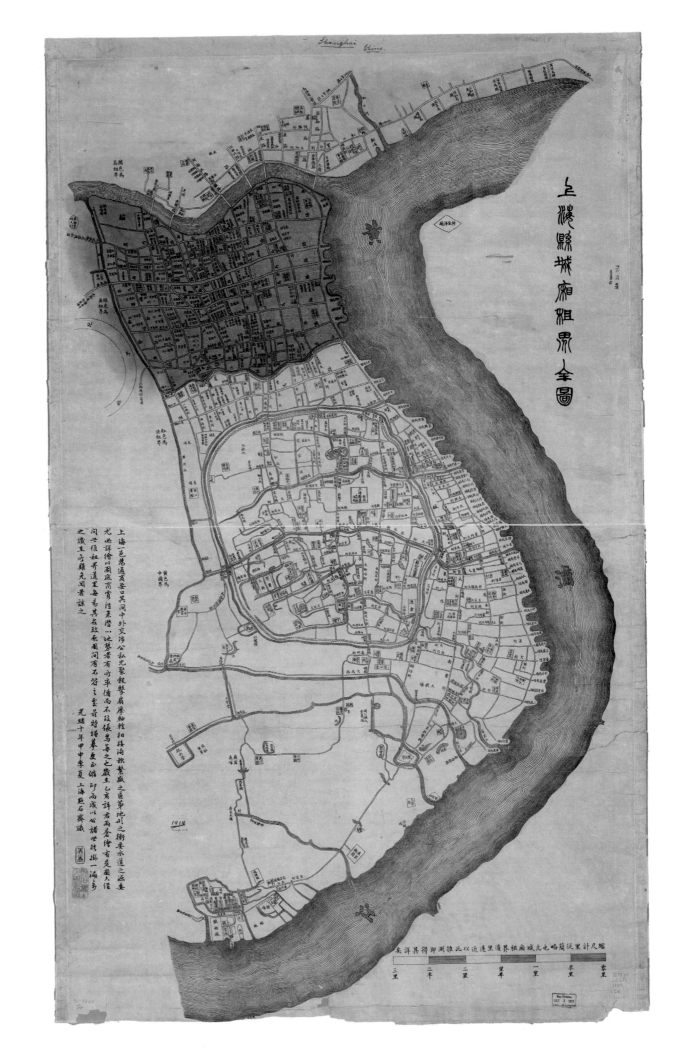

Hotbeds of innovation: 1800–1900

FOO-CHOW (FUZHOU), BY J. LESGASSE, 1884
Fuzhou was opened up to Westerners as one of the Treaty Ports after the First Opium War. Based on a survey in 1868, this cadastral map shows foreign property ownership in the Cangshan district to the north, facing Fuzhou (bottom right) on the Min River, where junks would anchor at the left. The area is connected to the south via an island and by the ancient Bridge of Ten Thousand Ages. The map identifies both residences and hongs, or trading factories, many of which were tea merchants. LEFT

SAN FRANCISCO, BY CHARLES R. PARSONS, 1878 In 1848 this was a settlement
with fewer than 100 buildings, but the discovery of gold turned San
Francisco into a flourishing city within a few decades. Seen from the
southwest, with the Pacific Ocean on the horizon, this original sketch
was published as a lithograph by Currier & Ives. The bay is depicted in a
panoramic sweep from Hunter's Point and Mission Bay on the left to
Fort Point, Golden Gate and Lime Point on the right. An earthquake and
fire in 1906 meant that rebuilding transformed the city, but the familiar
landmarks of Union Square and City Hall are there. ABOVE

VICE IN CHINATOWN, SAN FRANCISCO, 1885 Despite having encouraged the Chinese as immigrant labourers to build the railroads, from the 1870s onwards discriminatory laws were passed to try to push them out. Chinatown was depicted by officialdom as a den of depravity, criminality and insanitary conditions. Based on a survey of every building in 12 blocks, this map uses colour coding to denote their use: pink (gambling houses), yellow (opium 'resorts'), green (Chinese) or blue (white women) for prostitution, red ('joss houses' – temples) and orange (general Chinese occupancy). The map resembles those used to identify hotspots of epidemics, implying that the Chinese were a menace to society. **BELOW**

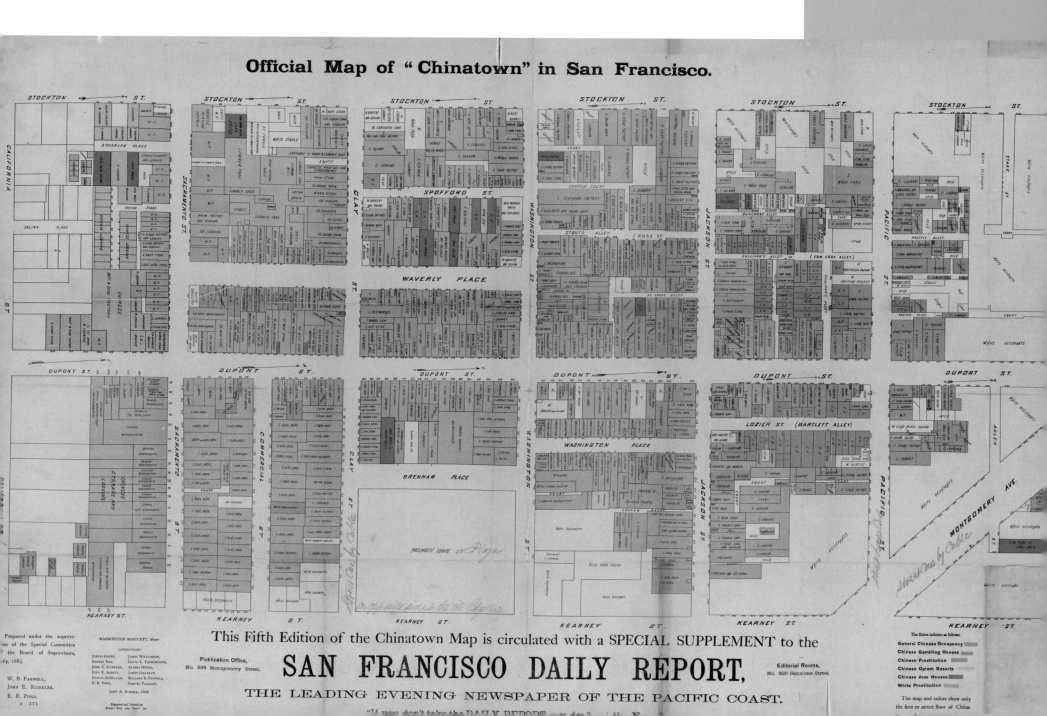

THE NATIONAL CAPITAL, WASHINGTON DC, 1884

This detailed view – 'sketched from nature by Adolph Sachse, 1883–1884' – is from east of the Capitol. The edition of the map bears the sponsorship mark of the B & O (Baltimore & Ohio) railroad, whose depot is shown just north of the Capitol. The map was published shortly before the Washington Memorial was completed. To the south of that obelisk today lies a Tidal Basin, on the other side of which is the Jefferson Memorial; west of both are more memorials to Lincoln, Franklin Delano Roosevelt, Martin Luther King, Jr, and the Vietnam Veterans. This emphasises the fact that the map shows the National Mall just two years after the US Army Corps of Engineers had begun a reclamation project for the tidal wetland known as the Potomac Flats, which continued until the 1890s. Dredging made the channel larger and more navigable for shipping, and it helped prevent flooding and the accumulation of sewage (flowing into the flats from the city's Washington Canal). Meanwhile the sediment it yielded was used to fill in the waterfront wetlands. The landfill formed East Potomac Park and extended the Mall beyond the site of the Washington Monument, adding more than 600 acres to Washington, DC. OPPOSITE

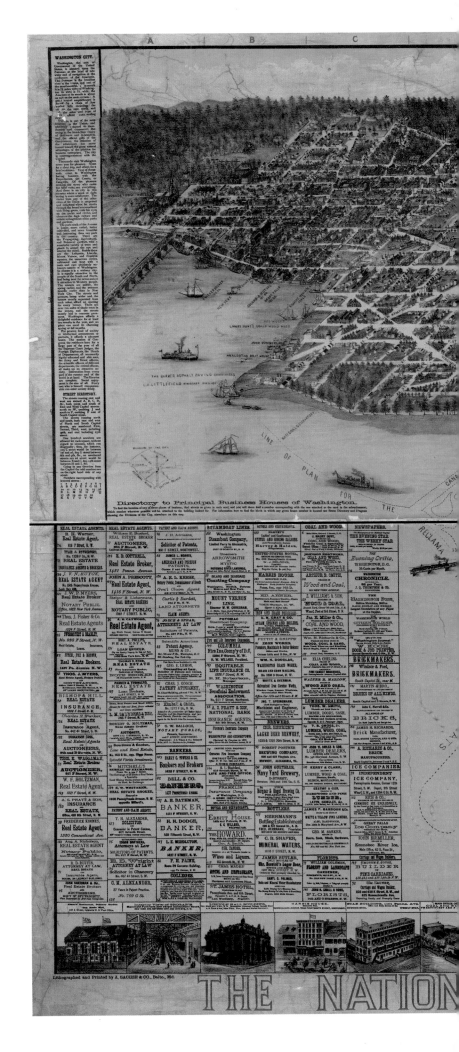

PICTURESQUE
B&O
ONLY LINE
VIA WASHINGTON.
ONLY LINE RUNNING 45 MINUTE TRAINS
BETWEEN
WASHINGTON
AND
BALTIMORE.

CAPITAL WASHINGTON CITY D. C.

Sketched from Nature by ADOLPH SACHSE, 1883-1884.

Copyright secured by A. SACHSE & CO.

BIRDS-EYE VIEW OF WASHINGTON CITY,
PUBLISHED AND FOR SALE BY
A. SACHSE & CO., Balto., Md.
Lithographers, Printers and Publishers,
ALSO, MANUFACTURERS OF
Metalic and Transparent Advertising Signs, &c.

Metropolis

CITY OF WASHINGTON, ANNUAL REPORT, 1880

The condition of Washington, DC, in the era after the US Civil War – with its dirt track roads and open sewers – was such that the District of Columbia embarked on a massive programme of infrastructure improvement through its Board of Works. In the 1870s, under Alexander Robey 'Boss' Shepherd, the Board filled in the canal, built hundreds of miles of paved roads and sidewalks, sewers, water and gas mains, installed streetlights, set up transportation systems and planted trees.

The effort continued for decades and during the late 19th century maps featured in an annual report, recording the progress being made. This selection of four includes the expansion of service provision, to provide streetcleaning, and a survey, presumably for taxation purposes, of property values, which are interesting to compare with services and infrastructure provision. Proximity to Pennsylvania Avenue, along which lay both the White House and the Capitol, in a central area then known as The Division, was expensive, while what is now desirable Georgetown was comparatively cheap. OPPOSITE

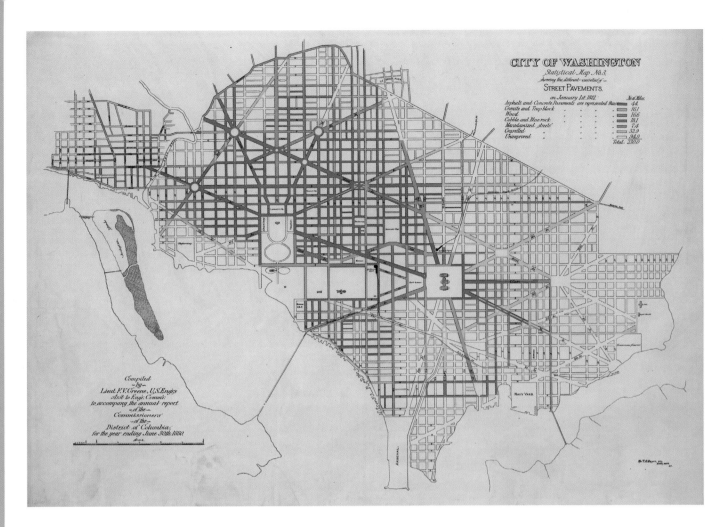

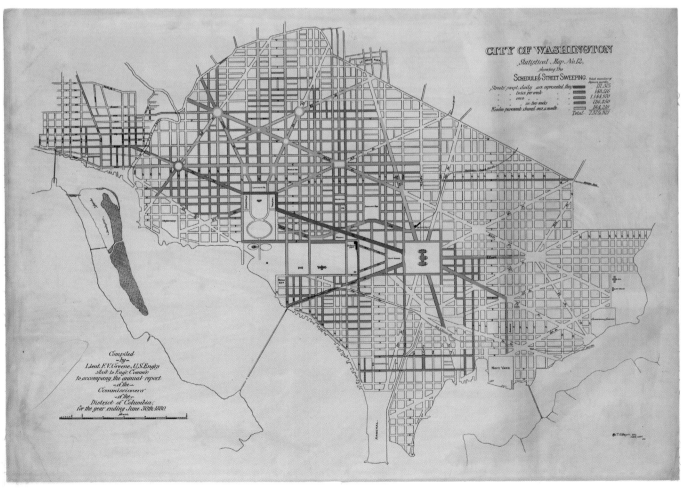

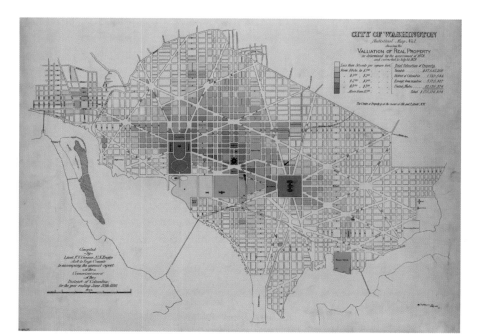

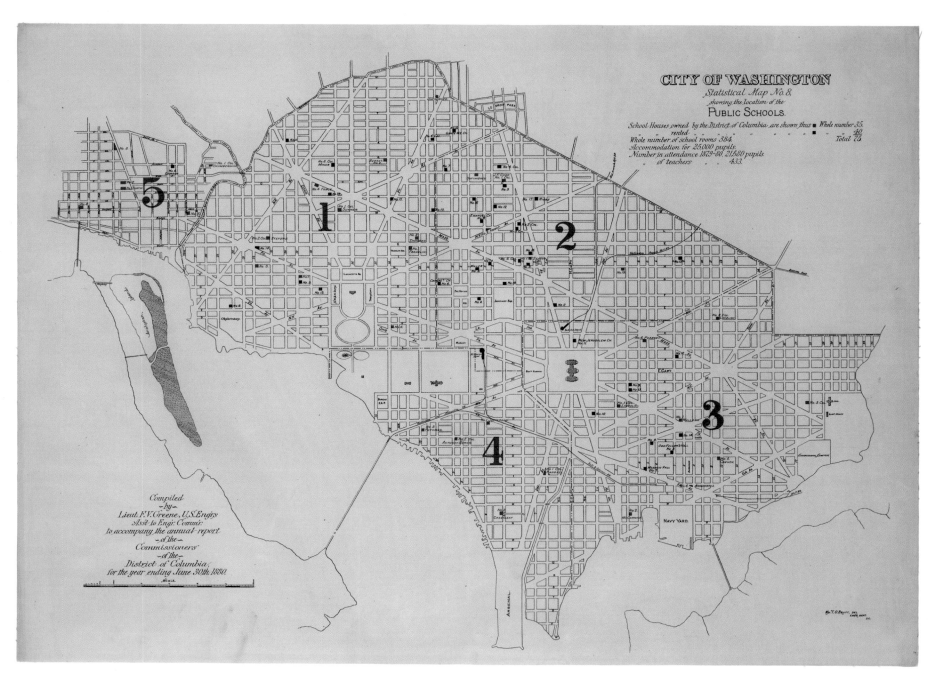

NEW YORK IN 1865 AND 1884. These bird's-eye views of New York City and Brooklyn both take a similar perspective from the south, with Battery Park in the foreground, the Jersey shore at top left and Williamsburg (Brooklyn) at bottom right. The 1865 image (left) was painted by John Bachmann, an important contributor to the development of bird's-eye views of the growing New York metropolis, while the later image (right) was created by Currier and Ives. The city's famous brownstone housing stock is already evident by 1865, but the standout sight is, of course, Brooklyn Bridge, opened in 1883.

'PICTORIAL ILLUSTRATION OF THE NIKKO MOUNTAINS', BY UEYAMA YAHEI, C.1880S

This is a stylised form of pocket map that lacks topographical accuracy. Nikko, a popular pilgrimage site in Tochigi Prefecture, has many temples, shrines and sacred spots, such as mountains, waterfalls and trees. Lake Chuzenji (top left) is particularly beautiful in the autumn. Japan's most lavish shrine is here: Toshogu (top right), which honours the founder of the shogunate, Tokugawa Ieyasu, whose grandson Iemitsu's mausoleum, Taiyuin, is nearby (top centre). Immediately beneath Toshogu in the map is the Rinnoji Buddhist temple complex. ABOVE

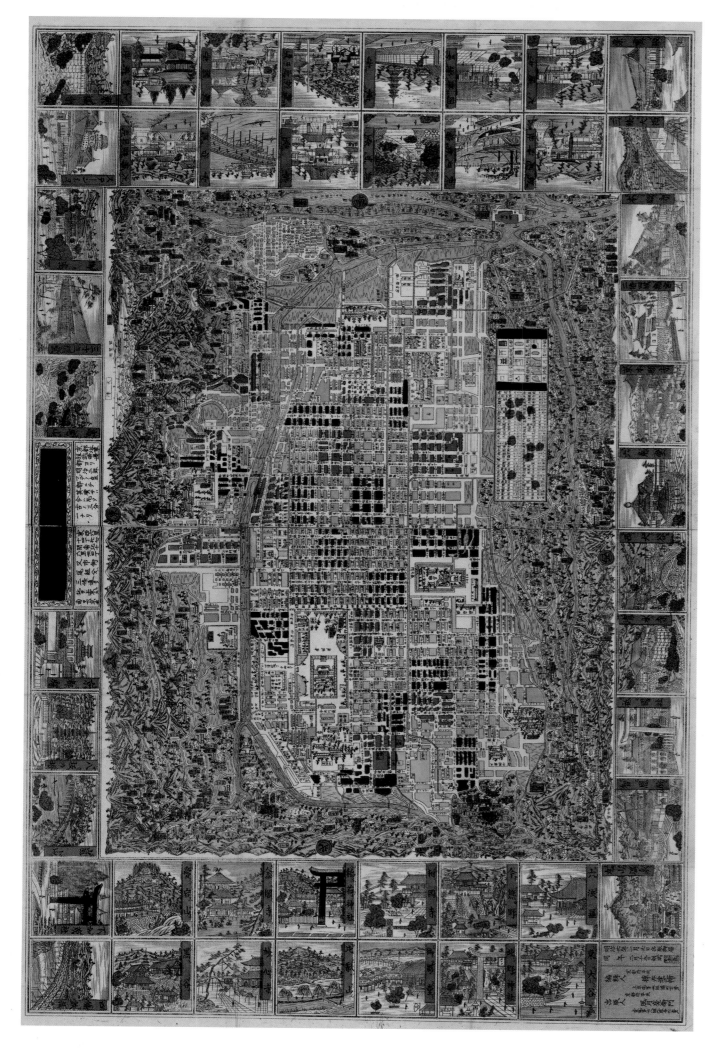

Hotbeds of innovation: 1800–1900

'KYOTO FAMOUS PLACES', BY TATSUNOSHIN KABAI AND SHOZAEMON KAZATSUKI, 1887 This map is an example of a genre that depicts the principal sights of Kyoto, dividing the city into colour-coded districts, with many locations illustrated in the border. The imperial palace complex is in the northern section at upper right, and Nijo Castle is in the centre and to the left. The area has approximately 2,000 temples and shrines, making it a trove of sacred places, ranging from small to monumental. In the same year as the map was created the Japanese government established the Kihinkai, or Welcome Society, aimed at encouraging foreign tourism. **LEFT**

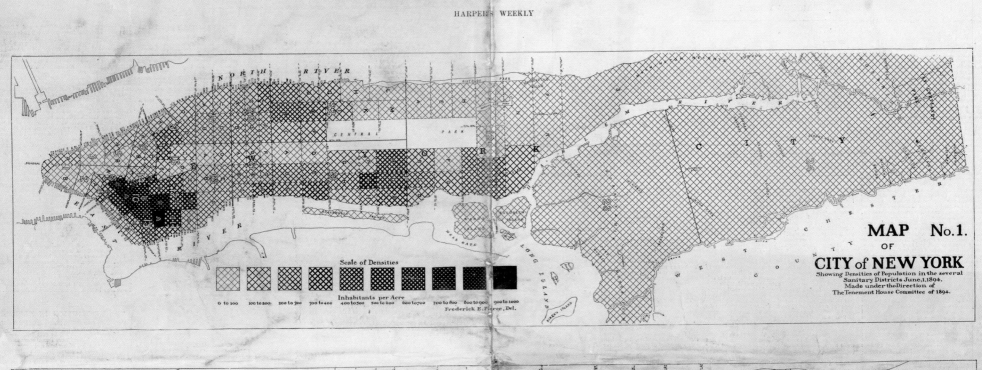

MAP No.1.
OF
CITY of NEW YORK
Showing Densities of Population in the several
Sanitary Districts June,1,1894.
Made under the Direction of
The Tenement House Committee of 1894.

Scale of Densities

Inhabitants per Acre

0 to 100   100 to 200   200 to 300   300 to 400   400 to 500   500 to 600   600 to 700   700 to 800   800 to 900   900 to 1000

Frederick E. Pierce, Del.

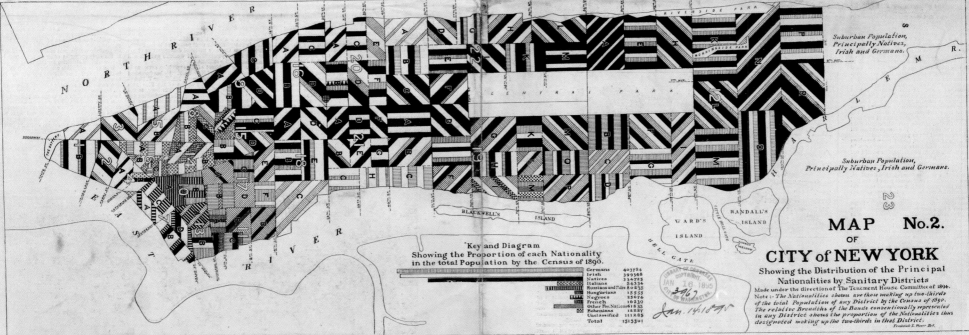

Key and Diagram
Showing the Proportion of each Nationality
in the total Population by the Census of 1890.

Germans	403784
Irish	399348
Natives	334733
Italians	54334
Russians and Poles	80835
Hungarians	15553
Negroes	25674
French	16239
Other Eu. Nations	61835
Bohemians	18887
Unclassified	111285
Total	1515301

MAP No.2.
OF
CITY of NEW YORK
Showing the Distribution of the Principal
Nationalities by Sanitary Districts
Made under the direction of The Tenement House Committee of 1894.
Note :- The Nationalities shown are those making up two-thirds
of the total Population of any District by the Census of 1890.
The relative Breadths of the Bands conventionally represented
in any District shews the proportion of the Nationalities thus
designated making up the two-thirds in that District.
Frederick E. Pierce, Del.

Suburban Population,
Principally Natives,
Irish and Germans.

Suburban Population,
Principally Natives, Irish and Germans.

THE TENEMENT-HOUSE COMMITTEE MAPS.—Drawn by Frederick E. Pierce.—[See Page 62.]

NEW YORK TENEMENT HOUSE COMMITTEE MAPS, BY FREDERICK E. PIERCE, 1894 Based on information from US Census reports, these maps present data in an innovative way that affords an at-a-glance appreciation of population density and nationality of origin. The maps first appeared in *Harper's Weekly* in January 1895, reflecting concerns in the city about tenement buildings and public sanitation issues – and, less overtly, ethnicity. It was claimed that the most densely populated place in the world was Sanitary District A of Ward 11, populated predominantly by German and Irish immigrants. **OPPOSITE**

BALTIMORE ELEVATED BUILDING MAP, BY J.T. LLOYD, 1894 The fine architecture of the city's business district was the subject of this local mapmaker, a choice which becomes poignant upon the discovery that within a decade many of these fine buildings had been lost. On 7–8 February 1904, a great fire began in the basement of the dry goods company Hurst & Co, located between Hopkins Place and Liberty Street on the south side of German Street, which is visible in the centre of the map. The greatest municipal disaster in the USA up to that time, the fire destroyed or damaged some 1,500 buildings in a 140-acre (57-hectare) area of the commercial downtown, disrupting thousands of businesses. **BELOW**

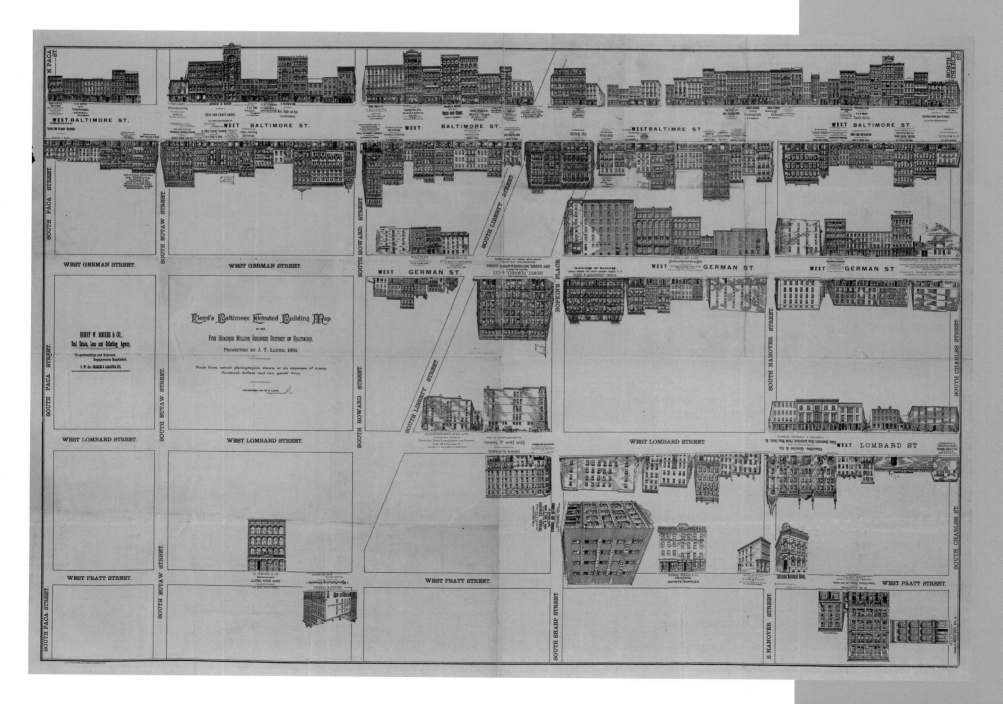

Metropolis

PLAN OF BARCELONA AND ITS SURROUNDINGS IN 1890, BY D.M. SERRA, 1891 This plan, prepared for the city council, gives an overview of Barcelona's development as a result of a rapid growth in population, which had changed in less than a century from around 115,000 in 1800 to 350,000 in 1880. In the 18th century the medieval city had eased overpopulation by reclaiming marshland and creating a housing development called La Barceloneta (the triangular spit of land in the old harbour at the bottom of this map).

The industrial age added greatly to pressures on space and in 1859 the walls were demolished to enable the city to expand into the hinterland. Engineer Ildefons Cerdà was commissioned to produce a plan by which this process could be managed. He provided for both public transportation networks and open spaces by using a geometric grid pattern in his scheme for Eixample Garden City. Eixample means 'extension' or 'expansion' and the area is readily identifiable beyond the dense old city and bordered to the north by the long thoroughfare of Calle Barcelona (now Avenida Diagonal). It was in this modern showcase district that Catalan Art Nouveau architects, such as Antoni Gaudí, were able to realise their designs, though many of the envisaged green spaces were not created. By 1890 the map reveals that development to the right of the northward artery of Passeig de Gràcia is greater than to the west.

RIGHT

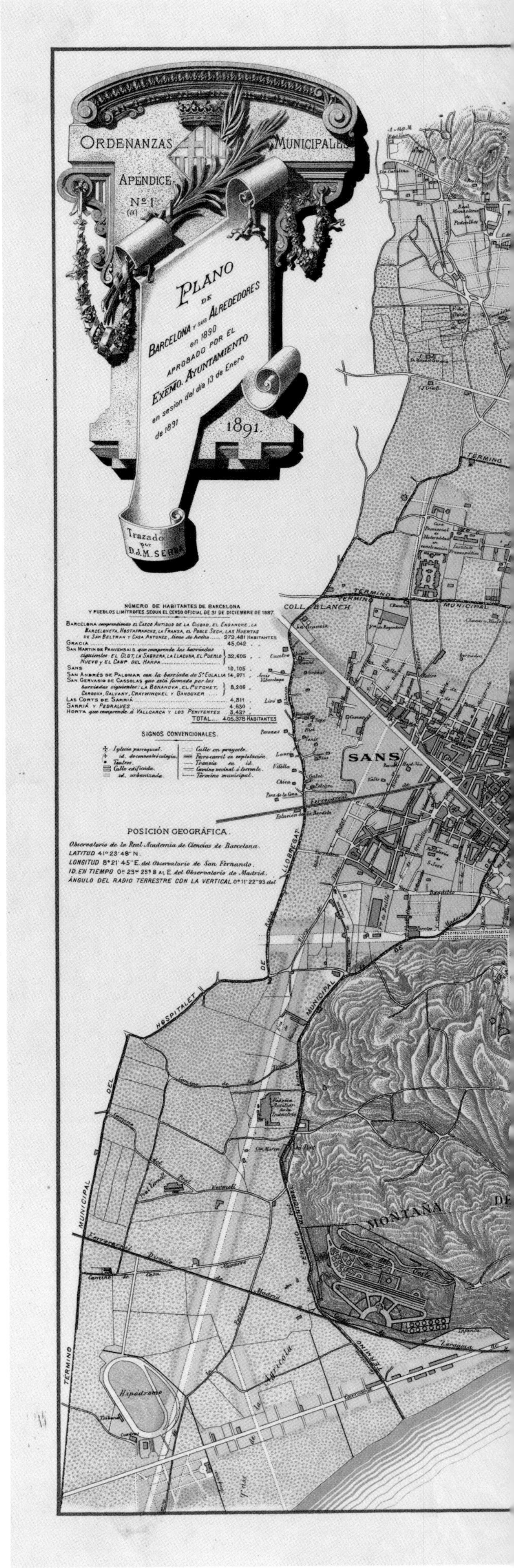

LOS ANGELES. CAL.

Population of City and Environs 65,000.

Published by SOUTHERN CALIFORNIA LAND CO., 344 N. Main Street. 1891

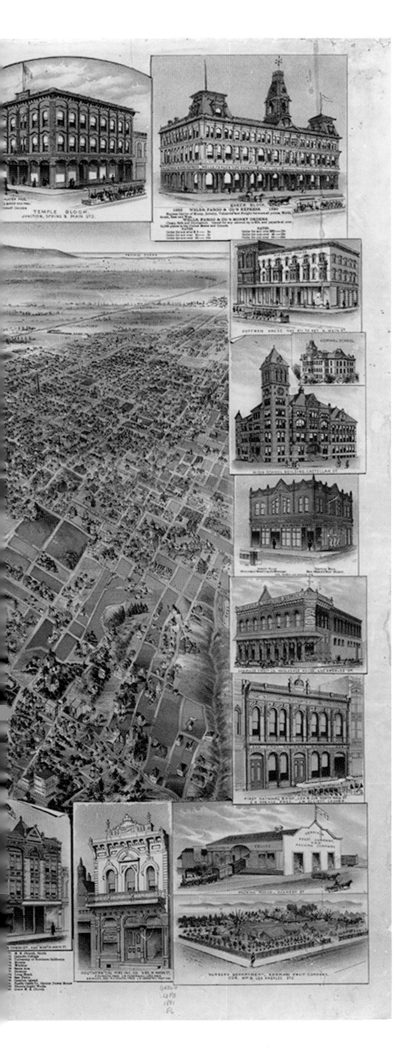

LOS ANGELES, BY H.B. ELLIOTT, 1891 This map – looking towards San Pedro Bay and Catalina Island (top, centre right) – was produced for the Southern California Land Company after a land-boom decade in which the city's population swelled from 11,000 in 1880 to more than 60,000. By 1900 it was more than 100,000.

The growth, and the city's sprawl, depended on infrastructure, such as bridges and roads (though the first car did not appear until 1897) to connect the city with outlying agricultural communities. Note G.J. Griffith's orange and walnut orchard in the map's central upper right. The first permanent bridge was built in 1870, linking LA with the east bank (Eastside) across the Los Angeles River. A covered, New England-style bridge, it can be seen at the end of Macy St next to the canning works (19 in the key, upper centre). The bridge helped the first two suburbs, known now as Lincoln Heights and Boyle Heights, to take root in the 1870s. The Macy St bridge lasted until 1904, when it was replaced by a structure felt to be more modern.

The atmospheric border depicts buildings such as City Hall and the County Court House. There is also the residence of Herman W Hellman, one of two Jewish brothers from Bavaria who were successful stationers, grocers and then bankers, after which they became major real estate developers in the city.

OPPOSITE

SATELLITE IMAGE OF LOS ANGELES, CALIFORNIA
The urban sprawl of Los Angeles, viewed from the air, reveals the contrast between pockets of extreme building density, intercut by highways, and the green of suburban areas. Satellites allow the constant changes in urban areas to be monitored with great accuracy.

PREVIOUS PAGES

The world's urban population grew even more rapidly in the 20th century than it had before and, to an increasing degree, nation after nation was acquiring an urban image and imagination. An 'urban' experience was increasingly one relating to cities rather than towns, and, more particularly, of really large cities. Consequently, direct experience of the countryside became relatively less common for people.

This shift to cities was worldwide. In 1900, large urban areas were concentrated in Europe and North America, with the largest being, in order of population, London, New York, Paris, Berlin, Chicago and Vienna. By that date, all the major American cities had been founded. During the 20th century, the urban areas in Europe and North America increased in extent, and there was also a marked shift in the location of the population from rural to urban; although this was already true in quite heavily urbanised countries such as England and Scotland, there was a significant migration from the land in France and Germany in the third quarter of the century as agricultural mechanisation gathered pace and as opportunities beckoned in areas that were expanding because of the industrial economy.

A WORLDWIDE TREND

The same process occurred in developing countries. In India, the urban percentage of the population rose from 10.9 in 1901 to 18.00 (19.0 on the 1901 criteria) in 1961, and that of Sri Lanka from 11.6 to 18.0 in the same period. Some of this was caused by the economic attractions (the pull factor) to cities which owed their existence to colonial European trading settlements, such as Bombay (modern-day Mumbai). Across the developing world, however, much of this shift occurred later: in China and Africa, for example, the trend was most evident in the last quarter of the

century. Although bad harvests might provide a specific incentive (or push factor) for this migration, more generally it was the case that rural life was often grindingly poor and the availability of jobs in towns encouraged migration, and life there had its appeal: towns and cities meant expanding horizons and they were centres of news, consumption, excitement and real, or apparent, social mobility.

By the end of the 20th century, the largest cities in the world were mostly in the developing world. They included Mumbai (Bombay, India), Lagos (Nigeria), São Paulo (Brazil), Dhaka (Bangladesh), Karachi (Pakistan), Mexico City and Shanghai (China). Lagos's growth reflected the movement of the population of Nigeria, the most populous country in Africa, towards the towns: the proportion of Nigerians living in an urban area rose from a fifth in 1963 to more than a third in 1991.

The process was repeated throughout the developing world. In South America, rapid population growth, a movement of people from the land, and the industrialisation of urban areas – not least because local production reduced the need for imports – ensured the steady upward curve of urban growth. Thus, the population of São Paulo rose from 1 million in 1930 to 17.1 million in 1990, that of Buenos Aires from 2 million to 12.6 million, that of Rio de Janeiro from 1.5 million to 11.2 million, and that of Lima from 250,000 inhabitants to 6.5 million.

Alongside pull factors, there were push ones. In the 1980s, the widespread droughts that affected the semi-arid Sahel belt in Africa led to many Mauritanians abandoning nomadic herding and moving to live in the city, especially the capital, Nouakchott. In other states, such as Angola, the disruptions of war accentuated this process, with Luanda growing rapidly. Conflict and floods led many Pakistanis to the cities,

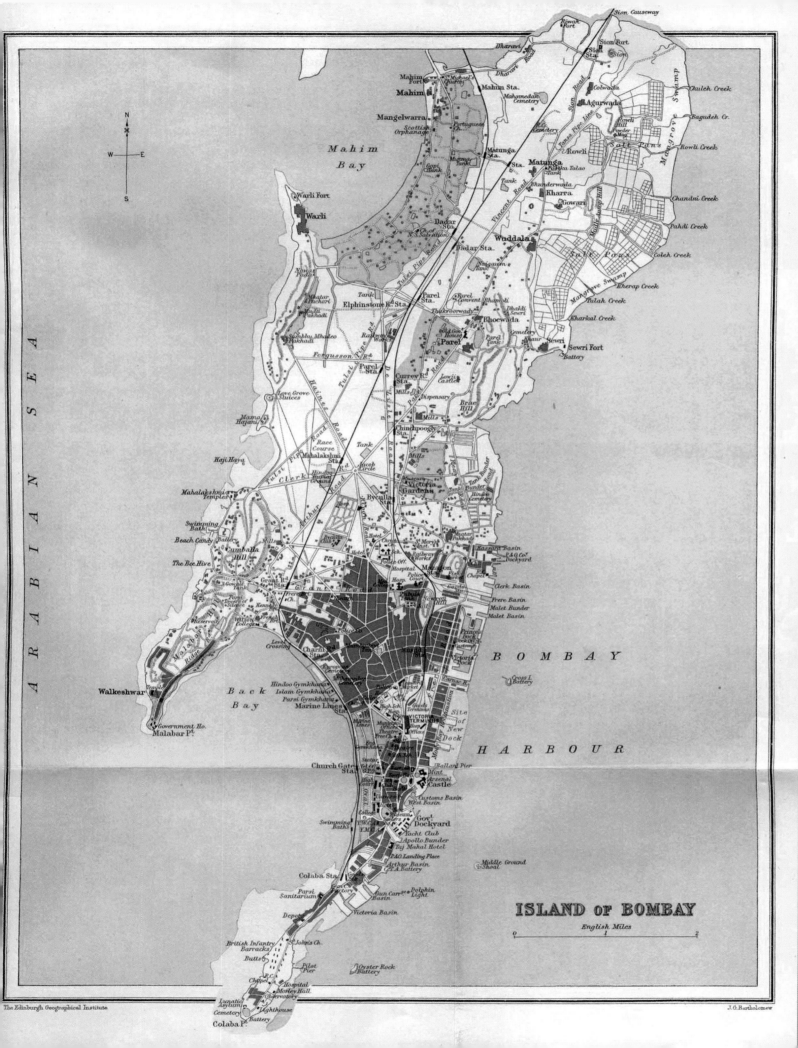

A global era: 1900–2000s

BOMBAY, FROM THE GAZETEER, BY S.M. EDWARDES, 1909 This map is from a three-volume gazetteer, *Bombay City and Island*, published in Edinburgh by J.G. Bartholomew. As well as this great trading city's hilly topography and green spaces, its infrastructure development (from railways to docks built and proposed) and the growth of settlement beyond the nucleus of the East India Company fort, dating from the 1720s, are both apparent. Perhaps most surprising is the fact that Bombay had originally consisted of seven main islands, which had become merged as one thanks to engineering reclamation work, much of it during the 1800s, that drained and filled in low-lying wetlands, some of which was paid for by the profits of the booming cotton industry. **LEFT**

ISLAND of BOMBAY

English Miles

The Edinburgh Geographical Institute

J.G. Bartholomew

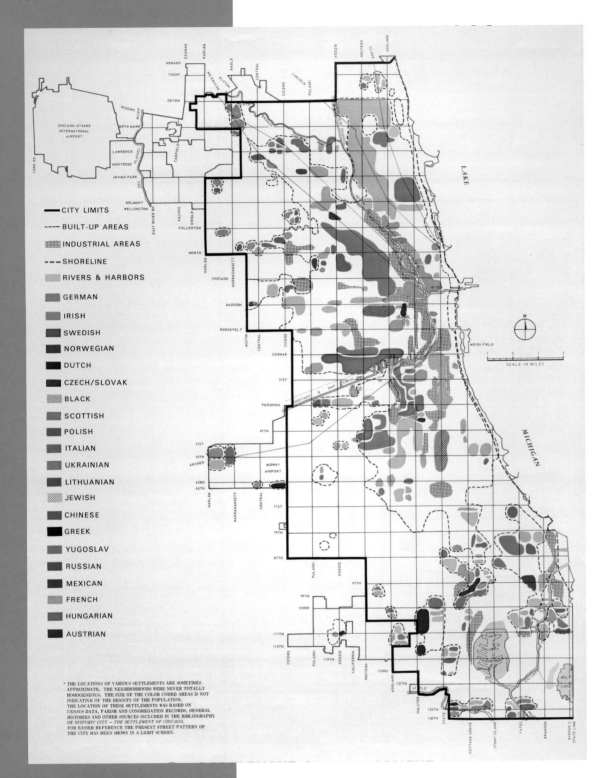

CITY LIMITS
BUILT-UP AREAS
INDUSTRIAL AREAS
SHORELINE
RIVERS & HARBORS
GERMAN
IRISH
SWEDISH
NORWEGIAN
DUTCH
CZECH/SLOVAK
BLACK
SCOTTISH
POLISH
ITALIAN
UKRAINIAN
LITHUANIAN
JEWISH
CHINESE
GREEK
YUGOSLAV
RUSSIAN
MEXICAN
FRENCH
HUNGARIAN
AUSTRIAN

* THE LOCATIONS OF VARIOUS SETTLEMENTS ARE SOMETIMES
APPROXIMATE. THE NEIGHBORHOODS WERE NEVER TOTALLY
HOMOGENEOUS. THE SIZE OF THE COLOR CODED AREAS IS NOT
INDICATIVE OF THE DENSITY OF THE POPULATION.
THE LOCATION OF THESE SETTLEMENTS WAS BASED ON
CENSUS DATA, PARISH AND CONGREGATION RECORDS, GENERAL
HISTORIES AND OTHER SOURCES INCLUDED IN THE BIBLIOGRAPHY
OF *HISTORIC CITY – THE SETTLEMENT OF CHICAGO*,
FOR EASIER REFERENCE THE PRESENT STREET PATTERN OF
THE CITY HAS BEEN SHOWN IN A LIGHT SCREEN.

because that had already occurred and national birth rates were significantly lower, but the nature of institutions was also an important factor. Areas of urban government or administration were often not amalgamated to reflect the spread of urban areas in cities such as London, Paris and Washington, DC. By contrast, in Canada in 1998 the suburbs were incorporated to create the Greater Toronto Area 'megacity'.

As people moved, over several post-war decades, to the lower-density housing in the suburbs, the loss of population from the traditional core of cities was to result in inner city decay in many parts of the developing world. This process was only reversed in the late 20th century, with many inner cities undergoing revitalisation and gentrification, often as derelict and abandoned former industrial areas were regenerated into residential districts. From Chicago and Baltimore to Liverpool and Barcelona, the redevelopment of urban industrial waterfront areas of wharves and warehouses into parks, offices, retail and leisure spaces has had a dramatic effect on many urban environments. This process accelerated in many places in the early 21st century. Maps of population density revealed these major growth trends in the 2000s in places as far apart as inner-city Manchester and Toronto.

Another factor affecting the stability or growth of populations in the cities of the developed world has been immigration, with newcomers from the developing world tending to concentrate in the major cities. In the 1950s, the Puerto Rican population of New York rose from 187,000 to 613,000. The arrival of Ukrainian refugees after the Second World War made Edmonton in Canada a key place in a Ukrainian Diaspora, while its larger counterpart among the Chinese was important in the expansion of both

notably Karachi. Elsewhere, refugees created shanty towns around cities such as Beirut.

MEGACITIES

In the second half of the 20th century the major cities of the developed world did not grow at the same rate as those in the developing world. In part, this was

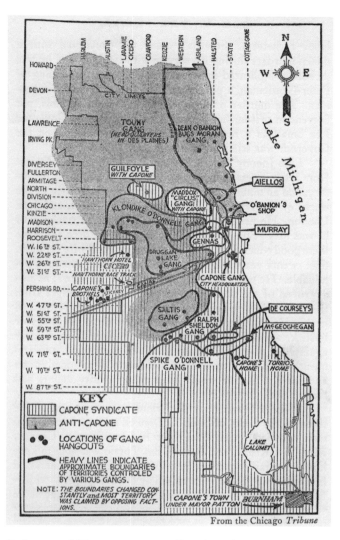

From the Chicago *Tribune*

Sydney and Vancouver, transforming them into more cosmopolitan cities.

URBAN CONCERNS: FROM SANITATION TO CRIME

Although several leading cities in the developing world were long established as major centres, none were prepared for the growth they experienced. The population of Bombay rose from 813,000 in 1901 to 4.15 million in 1961. Despite a marked increase in urban employment in the developing world, both in industry and in services, there were insufficient jobs to prevent a major increase in unemployment and under-employment.

The urban infrastructure proved particularly deficient on a number of counts, including water supply, public health, housing and transport. The percentage of the population in the developing world with access to safe drinking water and sanitation was greater in urban than rural areas, but, even so, many urban areas lacked clean water. This encouraged

epidemics of communicable diseases. Health provision was particularly poor for recent migrants into cities, many of whom were exposed to squalor, disease and great poverty, and lived in more marginal residential districts, especially squatter camps. Their illegal or semi-legal character ensured that these districts were not adequately represented in the mapping of cities. The proportion of the population who lived in such places reflected the inability of urban regions to meet demands for affordable housing.

Such urban areas proved difficult to police, and state authority in them was frequently limited, the nuances of which were difficult to map. In an echo of 1920s' Chicago, gangs competed with each other, and with the police, to dominate large tracts of cities such as Karachi and São Paulo, leading to high rates of urban violence. In greater São Paulo, the number of murders rose above 8,000 in 1998, with the rate of murders per 100,000 people passing 50 in 1999. Murder rates also climbed in other cities, from Johannesburg to Naples.

The development of cities is linked to another crime in the form of corruption, notably related land deals, as with Delhi's new airport, and planning permission. These are important issues, both when cities spread out into the surrounding countryside and when there are major changes within the existing urban area. Attempts were made to enhance value through changing land use. For example, in 2012 the pressure of development in Lagos – a city adding half a million people a year and currently heading towards a mid-century population of 40 million, with all the possibilities that created as a result of rising land values – led to attempts to cut down part of Makoko, a district that is built on stilts above the lagoon. This was an aspect of a more general attempt to develop land reclaimed from the sea, an engineering feat that has made a significant contribution to the

COMMUNITY SETTLEMENT MAP FOR 1900, CITY OF CHICAGO, 1976 This is one of six maps prepared for Commissioner Lewis W. Hill by the Department of Development and Planning to show variations in Chicago's ethnic distribution pattern for 1840, 1860, 1870, 1900, 1920 and 1950. The inclusion of the city's street pattern helps to locate neighbourhoods but recording ethnic information in a map is notoriously complex and a note on the map explains that areas of the city were never totally homogenous (not least because of intermarriage). Rather than being used for planning, publication in the bicentennial year of 1976 may indicate that Chicago was proclaiming its national importance and global heritage. OPPOSITE

CHICAGO'S GANGLAND MAPPED, 1925 Chicago had a long history of criminal activities such as gambling, extortion and prostitution, which tainted the city's politics and affected some of its ethnic groups. In the 1920s, during the turf wars between organised bootlegging gangs the term 'Public Enemy' was coined by the Chicago Crime Commission, established in 1919 by concerned citizens. The dominant position of Al Capone in Chicago's South Side emerges clearly in the map, as does that of Roger Touhy's gang on the Northwest side. The proximity of the various gangs helps explain why Chicago's so-called beer wars raged violently for several years and caused more than 200 murders. LEFT

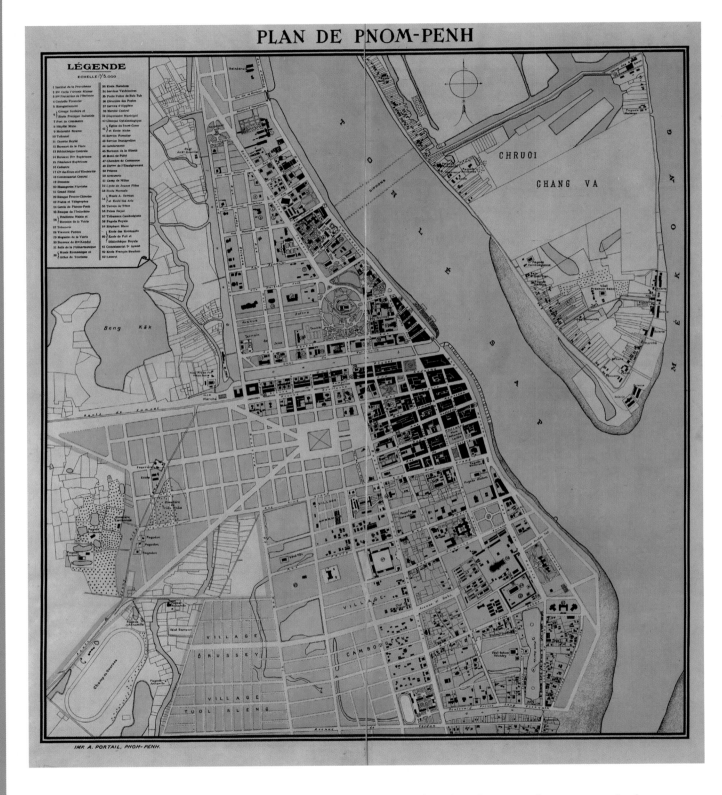

PHNOM-PENH, BY ALBERT PORTAIL PUBLISHING, 1920 The capital of Cambodia is a water-dominated city, sited at the confluence of the Mekong, Sap and Basaac (bottom right) rivers. From the 1860s, in the era of French control, Phnom Penh's traditional linear development along the riverbank and over the water, with stilted or floating buildings, was transformed. The city became an inland, gridded one of permanent structures perpendicular to the river. Drainage canals began the segregation of the city into ethnic quarters: the European one was bounded by the canals in the northeast (the city's Wat Phnom pagoda is in the southeast corner); the Chinese below the east–west canal; the Cambodian to the southeast. around the Royal Pagoda; and the Vietnamese to the west of that. RIGHT

consolidation and growth of many cities, including Bombay, Phnom Penh and Macau.

Lagos reflected the tensions of urban growth: many more people, terrible congestion, corruption, a poor power supply and a lawlessness that led to vigilante groups keeping the peace for payment. At the same time, attempts were made to deal with the crisis. These included, in the early 2010s, the construction (with Chinese help) of seven light railways and improvements in tax collection, which increased from

$8 million monthly in 1999 to $100 million in 2012. The tax revenue made it possible for Lagos to access capital and long-term debt markets, and to issue bonds.

URBAN AND NATIONAL IDENTITIES

With their greater proportion of the world's population, cities were at the forefront of political economic, social and cultural change. Following the First World War, British and French territorial power was at its height and cities under their control, from North Africa to Indochina, were remodelled to reflect this. In Morocco, Casablanca was redesigned in the 1910s–1920s and given wide boulevards as a centre for the new protectorate. There was a grand post office (1918), a *palais de justice* (1925), and a Catholic cathedral (1930). Other *villes nouvelle* (new cities) were established alongside traditional city centres.

After a long period of involvement in the region, the Dutch and British imperial powers were humiliated by the Japanese in 1941–1942 in the East Indies, Hong Kong and Singapore. From 1945 to 1975 European imperial power collapsed globally and political control over some of the world's largest cities changed as a result, with many becoming the capitals of newly independent states. Such changes affected not only those cities, but also wider understandings of urban identity, at least in a particular region. For example, in the Balkans and the Middle East, the sudden collapse in the 1910s–1920s of the longstanding Ottoman hegemony, the rise of multiple nationalisms, and the end of the cosmopolitan character of former multicultural port-cities, notably Izmir (formerly Smyrna) and Alexandria, led to major changes to the mental maps of wider patterns of identity and exchange. The rebuilding of Thessalonika by French city planner Ernest Hébrard, who later worked in French Indochina, accentuated a modern, Neo-Byzantine character rather than an Ottoman one.

Metropolis

**SAN ANTONIO, BEXAR COUNTY AND THE
SUBURBS, BY NICOLAS TENGG, 1924** The map
shows the subdivisions and house
numbering systems of a city that has
been French, Spanish, Mexican, Texan and
Confederate, but is now part of the USA.
Although first settled in 1718 as a mission,
near the headwaters of the San Antonio
River, the development and expansion of
San Antonio de Béxar, was quite slow
until the city entered the Union.
Developed from the 1870s, the Fort Sam
Houston complex of military buildings
dates from that era. By the early 1900s,
San Antonio was the largest city in Texas
(today it is second to Houston). As the
city has grown it has successfully blended
in the old, such as the Spanish mission
better known as the Alamo, but the most
celebrated part of San Antonio's urban
landscape is the river that snakes through
its heart, which has been developed as
the Paseo del Rio, or riverwalk. **RIGHT**

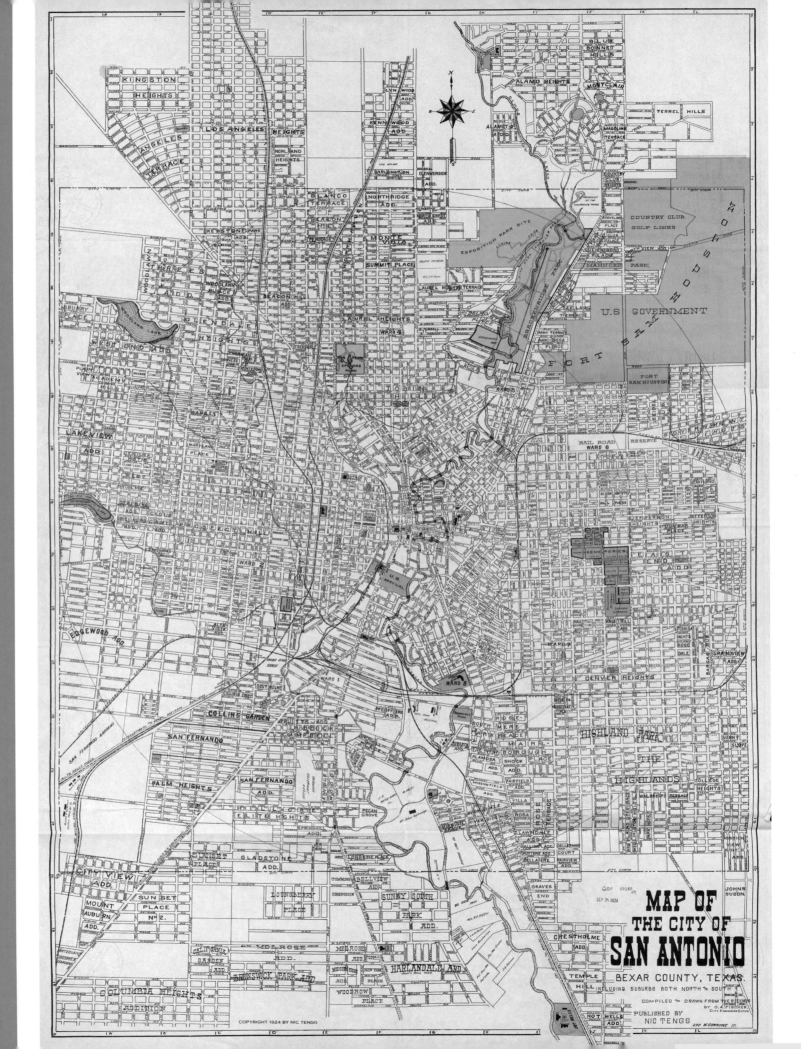

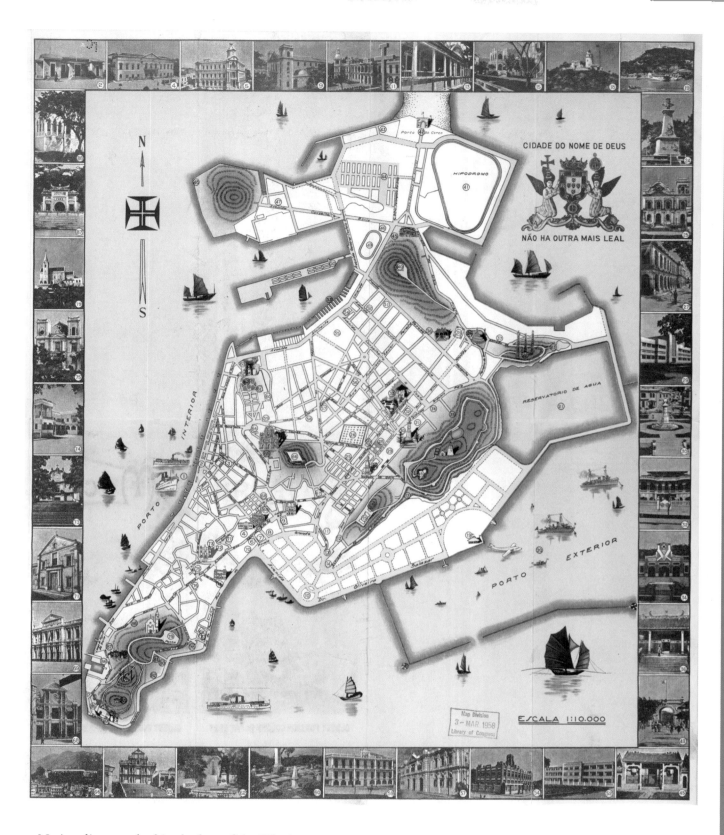

CIDADE DO NOME DE DEUS

NÃO HA OUTRA MAIS LEAL

ESCALA 1:10.000

MACAO TOURISM MAP, BY HONG KONG PRINTING PRESS, 1936 Founded in 1557, Macao (Macau) became the principal port for Western trade with China in the 17th and 18th centuries. A governor of Hong Kong, Sir John Bowring, called it the 'Gem of the Orient Earth'. Reflecting its atttraction to outsiders, this map is part of a detailed foldout produced by the Macau Tourism Agency. The map is bordered by 36 photos of the 88 places listed and numbered in the accompanying key. The city extends from Barra Point and the fortress of Barra at the southern tip to the northern boundary marked by the Barrier Gate, flanked by firecracker factories, houses for the poor and the racecourse (a telltale European creation). The new harbour is on the east (Porto Exterior), above which is the city's fresh water reservoir, and the wharves for the Hong Kong and Canton steamers are on the west side. **LEFT**

Nationalism resulted in the loss of the West's favoured-nation status in some Chinese cities in the 1920s, a process that continued with the Japanese attacks in the 1930s and 1940s, and the triumph of Communism from the late 1940s. In 1997 Hong Kong was returned by Britain and in 1999 Macao by Portugal. The net effect was a loss of multicultural identity.

THE NEED FOR RESIDENTIAL SPACES

Maps, unless they were thematic, often provided scant intimation of the nature of the urban experience. This was certainly true of housing in the developed world. Municipal building programmes ensured that a large percentage of the population lived in publicly provided housing. The demand for new housing was such that

KOBE: MUNICIPAL WATER SUPPLY DISTRIBUTION,
FEBRUARY 1945 During the Second World
War Kobe was Japan's largest port, an
important industrial centre, and its
sixth-largest city. This map was produced
by the US Office of Strategic Services,
Research and Analysis Branch, and it
records Kobe's inadequate water-supply
infrastructure. Although there had been
major work before the Second World
War , demand from population growth
strained capacity. The first filtration plant
at Okuhirano, completed in 1900, was
only the seventh modern waterworks in
Japan. Karasuhara dam was completed in
1905 and Egeyama in 1926, but more were
needed and air raid damage resul ted in
massive leakage rates. Japanese cities,
with many wooden structures, were
particularly vulnerable to firebombing
and more than half of Kobe was lost to
this often catastrophic form of aerial
attack. **RIGHT**

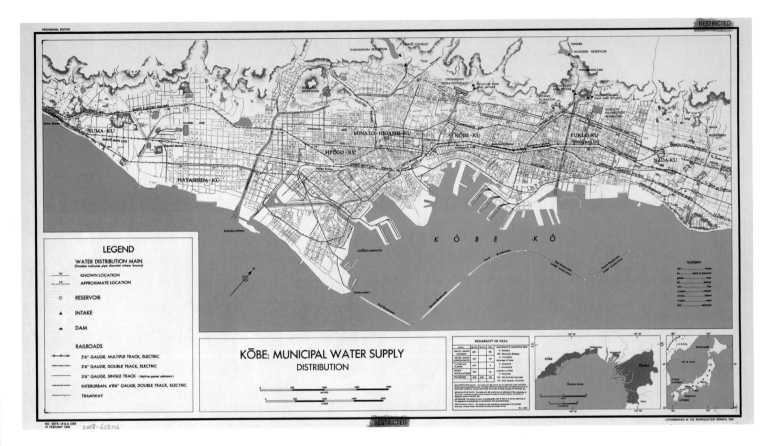

private landlords were unable to satisfy it, and changes
in tax policy and investment during the first three-
quarters of the century only encouraged this decline of
private rental housing. Public provision had to increase
to compensate, although in the USA the private rental
sector remained more important than it did in Europe.

Throughout the developed world, the number
of city-dwellers owning their own homes rose, but this
remained an issue determined by social class, with the
poor, most of the working class, and much of the
middle class unable to join in this process of
ownership. This social configuration of housing
varied greatly by country.

Just as housing configurations differed worldwide,
contrasting political developments ensured that other
city-related activity produced a patchwork of patterns.
For example, the response to the car varied greatly: in
China, under Communist rule there was authoritarian
state control, including opposition to private transport;
whereas in many Western cities, urban and suburban
development was obliterated to make way for major
highways carrying motor vehicles.

From the 1950s onwards the Communist
government in China made a major attempt to provide

housing by throwing up, as in the West, system-built
blocks of flats, and this approach remains the case
today. In the process, older vernacular cityscapes have
been destroyed without consultation, most notably in
the run-up to the 2008 Olympics, when many of the
traditional Beijing neighbourhoods known as *hutongs*
were bulldozed. With their narrow alleyways and
courtyards, these cityscapes did not lend themselves to
the scrutiny of government and the streamlined clarity
of planners – a clarity captured in urban maps when,
for example, some European cities are compared pre-
and post 19th-century modernisation. The destruction
in Beijing, Shanghai and elsewhere underline the
extent to which new cityscapes are not simply about
housing people but also represent political and
economic power and cultural assumptions.

At least in China, there was a major effort made to
provide housing. In contrast, in much of the
developing world the state made little or no effort to
meet demands for housing, nor to regulate its quality.
In some countries, there have been more serious
efforts albeit generally inadequate. Thus, in 2003
suicide bombings in Casablanca led to an attempt to
tackle the problem of the city's vast shanty towns, by

providing new housing. In practice, the number of houses being built could not keep up with the pressure of population growth and migration from the countryside.

STYLES OF DEVELOPMENT

China exemplifies a more general point that, rather than pretending that change is uncontroversial, it is important both to note controversies and to explain how these affected the process of development. In the West, the suburbia of the 1930s and the building that followed the Second World War reflected contemporary aspirations, only for their effects to become controversial. To critics, suburbia was simply sprawl, with open space having been built over and a culture created that was heavily reliant on the car.

This concern encouraged higher-density housing after 1945, a development to which both high land prices and advances in construction technology contributed – as did a fashion factor, whether architectural or planning. Communal architecture by people like Le Corbusier (the nickname of Charles-Edouard Jeanneret-Gris), most influentially his *Unité d'habitation* estate in Marseilles (1945–1950), proved that prefabricated methods of construction enabled large-scale multi-storey blocks of flats and office buildings to be built rapidly and inexpensively. This form of development was seen across the developed world.

Even more were planned: in Havana, José Luis Sert produced a set of plans to destroy much of Old Havana and replace it with tall concrete and glass buildings, although his schemes were never implemented.

It is difficult to recover the attitudes of the 1950s. Many of the buildings of the time were generous with space, and the quality of the urban environment is better glimpsed from detailed plans rather than maps. The new buildings of the 1950s often provided people with their first bathrooms and inside toilets. The weight of evidence shows that the problems of poorly built estates were more apparent from the 1960s, whereas earlier examples were frequently better built and more popular.

From the 1960s onwards, municipal multi-storey flats ('projects' in the USA) were often criticised for being of poor quality, ugly, out of keeping with the urban fabric, discouraging community feeling, and breeding alienation and crime. Housing such as the Heygate and Aylesbury estates in south London offered a dispiriting urban environment, but mapping cannot really reflect this. Similarly problematic is presenting the character and impact of social and ethnic groups. Parcelling residential areas into discrete 'communities', in the style adapted from the traditional urban sociology of the Chicago School, helped the cartographer, but it is an approach fraught with difficulty, not least because it suggests to the viewer clashing regions of homogeneity that belies a more complex reality.

Issues are often more complicated and nuanced than they appear. The issue of housing provision or community identities are affected by poverty, which removes the element of choice. That mass-produced public housing is not inevitably of low quality has been shown by Scandinavian cities such as Stockholm. The failure of many tower-block estates, designed as communities with elevated walkways called 'streets in the sky', may owe more to an absence of social cohesion rather than poor planning. For example, the high-rise new towns built in the New Territories of Hong Kong from the 1970s, such as Sha Tin, avoided many of the social problems seen in Britain in schemes of this kind.

Despite these issues, and again poorly recorded in maps, during the century housing standards improved across the developed world, and houses were integrated into wider networks, with the provision of electricity, gas and telephone. None of this was true for much of the developing world; instead, urbanisation there saw a borrowing of international forms of housing construction, but with only a limited infrastructure provision and generally inadequate standards.

SHIFTING CENTRES OF GRAVITY

Modernist skyscrapers proved more successful as an international style for city-centre office buildings and hotels. These skyscrapers proclaim a city's status – as in Hong Kong with the juxtaposed buildings of the Hongkong and Shanghai Banking Corporation (designed by Norman Foster), the Bank of China Tower (I.M. Pei) and the Standard Chartered Bank. Such buildings were products of the financial liberalism and easy credit of the 1980s and early 2000s. However, aside from problems with ambitious projects when fiscal circumstances got worse, as in the late 1990s in Asia, such environments depend on plentiful electricity supplies. An interruption can create problems such as the 2003 Great Blackout in North America (in the northeast USA–Canada border area) or in Washington, DC, in 2012.

Within countries, there were shifts in the relative influence of cities. In the USA, the prime setter of the urban model, the dominance of national life by the major cities of the East and Midwest, remained central during the first four decades of the 20th century, not least because they lay behind much of its industrial growth, notably in the car industry. However, Second World War industrial activity shifted economic activity to the West Coast.

Moreover, the economic and demographic significance of the West, notably of California and Seattle, became more important in the subsequent 'baby boom', the rapid population growth of the late 1940s–1950s. There were also spectacular developments in the cities of the 'New South', especially Atlanta, Dallas and Houston.

This shift challenged the cultural influence of the East and, particularly, of New York. Indeed, there was a growing assertion on the part of regional challengers, most dramatically in the 1950s with the Beat Movement that emerged in San Francisco. Alongside this regional assertion there was a change in the national cultural focus. Unlike the film industry, television had remained more New York-oriented, but in 1972 NBC's *The Tonight Show*, presented by Johnny Carson, moved to Burbank in the Los Angeles conurbation.

Between the 1950s and the 1970s, there were major changes in the USA in the distribution of manufacturing, moving from the Northeast and the Midwest, to the so-called Sunbelt of the West and Southwest, as well as into parts of the South. The resulting sense of opportunity produced substantial population changes.

From 1960, the population of the greater Phoenix area grew by an average of 47 per cent per decade, so that in 2003 Phoenix displaced Philadelphia as the USA's fifth most populous city. This was part of a major reordering in the urban hierarchy. Between the censuses of 1990 and 2000, Detroit lost 7.5 per cent of its population and Pittsburgh, as a steel city and traditional industrial centre, lost 9.6 per cent. In contrast, the success of high-technology industries contributed to the population of Austin, Texas, growing by 63 per cent between 1990 and 2003, to 1.4 million.

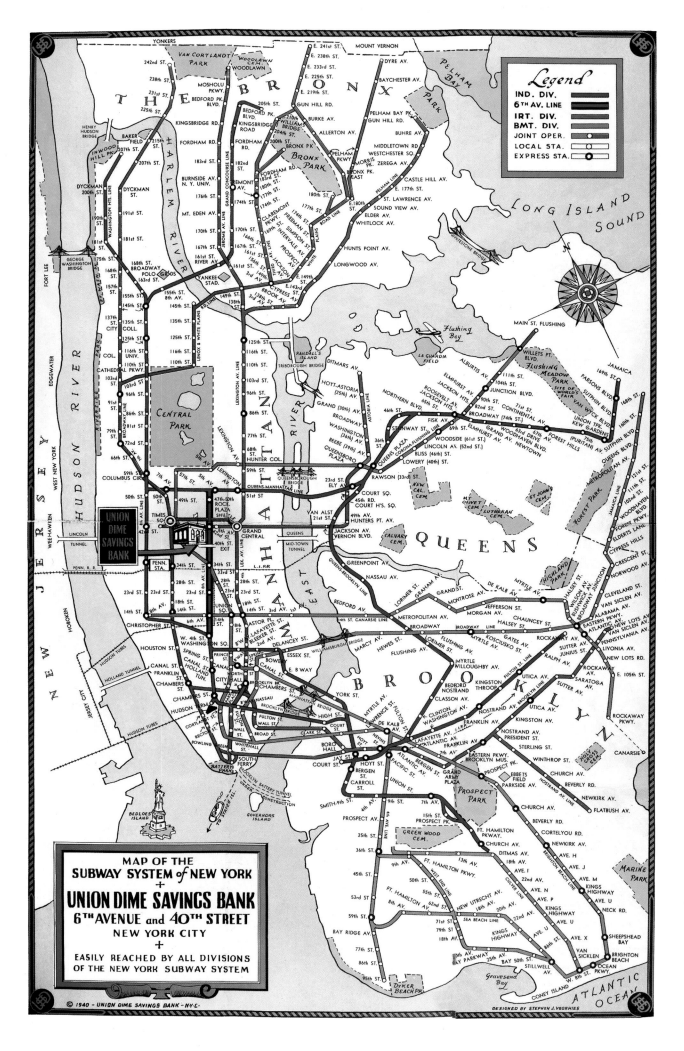

MAP OF THE SUBWAY SYSTEM OF NEW YORK, BY STEPHEN J. VOORHIES, 1940 This is the first of seven editions of this map, produced to the early 1960s by the Union Dime Savings Bank in New York, which it described as 'a community service for New Yorkers but also for the assistance it may lend to the thousands of visitors who daily come to this great city'.

Established in 1859, by 1876 the bank had bought a property on 32nd and Broadway in the 34th Street district, which it sold 34 years later for a record price thanks to the transformation of the area's fortunes into a retail hub caused by the arrival of stores such as Macy's and Gimbels, and new transportation links, such as Pennsylvania Station, to add to the 6th Avenue elevated train that had serviced the 34th Street area since the late 1870s. The bank used the sale proceeds to build an Italian Renaissance-style edifice on Avenue of the Americas (6th Avenue) at 40th Street. **LEFT**

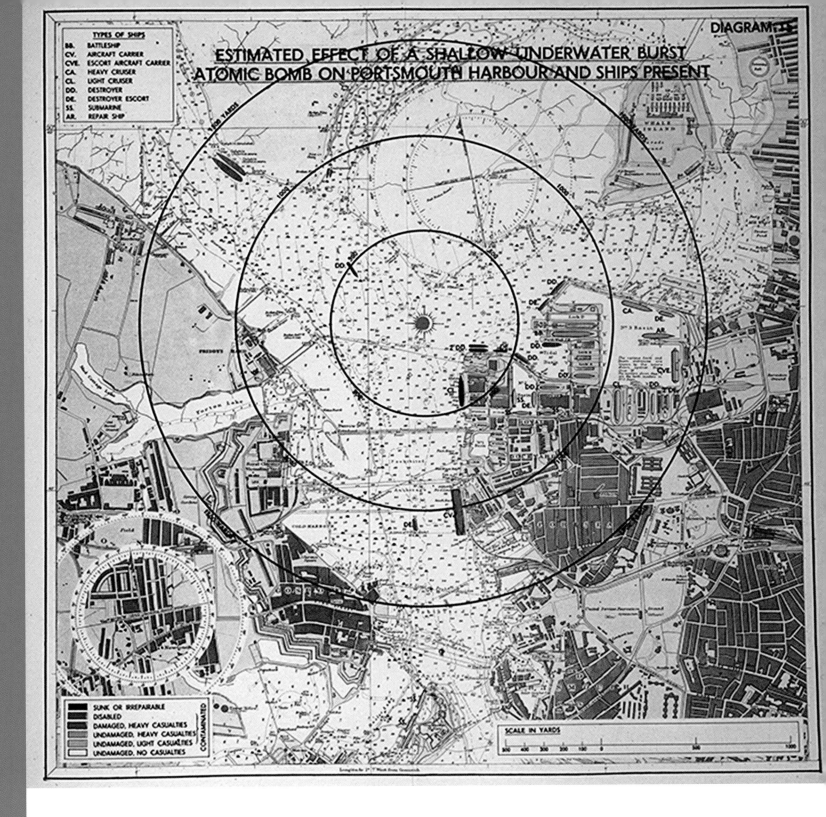

THE ESTIMATED EFFECT OF A NUCLEAR BOMB ON PORTSMOUTH NAVAL DOCKYARD, 1947 The strategic bombing campaigns during the Second World War and then the development of nuclear weapons made cities vulnerable to a level of destruction not envisaged in previous eras. This map reflects lessons learned by British observers at Operation Crossroads in 1946–1947, when two atomic bomb tests (air and underwater detonations) and subsequent surveys were conducted by the USA at Bikini Atoll in the Pacific. The onset of the Cold War meant that the threat of Soviet nuclear attack became a major issue for British defence planning. **RIGHT**

CHANGING CITYSCAPES

Moreover, the shape of cities, notably but not exclusively American ones, changed as a result of suburban expansion. The spreading use of cars; decentralisation in employment; increasing leisure, education and other service activities; inexpensive new-build construction methods; and the ready availability of mortgage support all helped to make suburban expansion profitable and possible. In place of 19th-century factories that had large labour forces, most modern American industrial concerns were capital intensive and employed less labour. They were located away from the central areas of cities, on flat and relatively open sites with good road links. There was a comparable shift in docklands, away from city waterfront anchorages, for example in New York and

San Francisco, to large, new 'greenfield' container ports that employed far less labour than the traditional, by now heavily unionised, docks. Similarly, wholesale markets in city centres were suburbanised. In 1969, Paris's markets were moved from the central site of Les Halles (where there had been a market in 1135) to Rungis, and the vacated location was replaced by a modern shopping centre.

In addition, the shift from manufacturing to the service sector meant a greater focus on the suburbs. By 2004, suburbia accounted for about 90 per cent of new office building in the USA. The spread of the suburbs provided many problems for the provision of public services. In greatly expanding urban areas in the USA, such as Atlanta, Houston and Los Angeles, far-flung water and sewage services had to be established as the suburbs spread, creating new 'edge-cities'.

The desire for space contributed to the spread of suburbia everywhere, as did rising numbers and improved communications. Lifestyle issues were also important, with people in many countries expressing a preference for developments of houses with gardens instead of city-centre flats. Thus, in the northern suburbs outside Madrid in the 1980s and 1990s places such as Puerta de Hierro arose.

NEW TECHNOLOGIES

Meanwhile, technological change made it easier to produce new maps at a reasonable cost. In turn, the look and feel of maps changed with the digitalisation of data and digital maps, which made it simple to produce maps in which it was possible to vary scale, projection and perspective, thus providing an image of the city that could be crafted for the individual viewer. Satellite navigation (satnav) systems were one of the consequences of technological innovation, and the difference between A–Z maps and satnavs reflected the

impact of changes in mapmaking technology and the impression of the city that results.

Both types of map were a product of a prime need for users of urban maps: searching for locations and routes. The need for the search reflected the increased role of the car in transport provision. With trains, the route was not the responsibility of the traveller; the eventual destination determines the choice of ticket. With cars, the route is the responsibility of the traveller and more information is required, notably the variety of road options.

These map systems were the principal ones in use, and the impression of the urban system that they provide is different to a heroic account of the grandeur and scale of the city as a whole. This difference was related to the fracturing of the reality and image of the city as a result of the extent and nature of suburbanisation. In particular, many residents came to know little about much of their city. This was a product not just of the scale of cities and the extent of personal transport use, but also of the development of suburbs as distinct entities with particular characters that were often different to those of the inner city or of other suburbs. Although this process did not begin in the 20th century, it became far more pronounced during that period. Growing up in London from the mid-1950s to early 1970s, I eventually knew Northwest London well, not least the tranche from there into the centre, but I really did not know South or East London.

As a consequence of such experiences and needs, the significance of maps of the entire urban area became less clear. And therefore also for tourists and other travellers who tended to spend their time in the centre. The production of 'mental' maps, notably the Manhattanite's view of the world, adopted a satirical approach to the topic, but it was one that captured the

fundamental importance of such a perception. The
nature of mental maps overlapped with other
depictions that were more ostensibly pictorial, such as
night-time Berlin in Conrad Felixmüller's 1925
painting *The Death of the Poet Walter Rheiner.*

DATA AND DISCONTENT

In response to the scale and complexity of cities, it was
necessary to develop new techniques to depict them.
The conventional procedure of equal-area maps risked
making the central area too small in order to provide
pro-rata coverage for the suburbs. Many maps
therefore followed the practice adopted by the London
Underground from the 1930s, which treated the
suburban areas at a less generous scale. The adoption of
a diagrammatic format conceals that process but makes
for a more practical map for the user.

The iconic London tube map is an example of the
thematic mapping that was increasingly to the fore in
city (and other) mapping. Although this mapping had
developed in the previous century, it became more
varied and popular as the availability of data increased,
as data was increasingly presented with spatial
indicators that could be mapped, and as computing
power made it far easier to process and present large
quantities of data. For example, the ethnic character of
city demographics could be better analysed and
mapped, and social information of the type that
Charles Booth had accumulated, often linked to the
ownership and character of housing, has become vast.

Other forms of thematic mapping were less
welcome, such as the city being a target for military
operations. Weapons technology that made air and
missile attacks, or the threats of such attacks, against

cities possible reduced the distinction between front line and home front. Beginning in the First World War and fully developed during the Second World War, the aerial bombardment of cities killed hundreds of thousands of civilians, who had the misfortune to live in places deemed to be legitimate targets of war.

Another key aspect of such a development arose from the extent to which cities became major sites for ground conflict, notably in the insurrectionary warfare that became more common in the second half of the 20th century. In such instances, mapping served not only for the planning of FIBUA ('fighting in built-up areas') operations – for example, the Soviet suppression of the Hungarian rising in Budapest in 1956, but also in depicting the scale of opposition. The latter served to suggest the inevitability of change in Cairo, Tripoli, Damascus and Aleppo, key centres of the 'Arab

Spring', during 2011–2012. In turn, the urban fabric of such cities posed problems for anyone seeking to maintain control militarily. Tanks can move along wide boulevards, but complex warrens of narrow streets, such as exist in old cities like Aleppo, pose serious problems.

The suppression of popular protests in Kuala Lumpur (1969), Casablanca (1981) and Beijing (1989) showed that in the 20th century cities had lost none of their potential as epicentres of opposition to established order, and that governments cede control of them at their peril. Demonstrations in key cities in the 1980s did much to overthrow Communist rule in Europe, including Bucharest in 1989. Cities, our greatest modern centres of focus for public attention, undoubtedly remain the crucibles of social change in our overwhelmingly urban world.

BERLIN, CITY OF DIVERSITY, 2012 To celebrate the 775th anniversary of Berlin in 2012, Kulturprojekte Berlin created a large-scale map (at a ratio of 1:775) in the city's Schlossplatz. The theme of the map, navigated using named main streets, is the contribution made to the city's historical development by immigrants, from early settlers, to the Huguenots to the modern-day creatives and entrepreneurs. At 775 locations a pole marks a place connected to an individual. For example, the city's Berliner Weisse beer was created by Swiss immigrant Daniel Josty who set up a brewery near Prenzlauer Tor in 1820, where Soho House hotel now stands. **ABOVE**

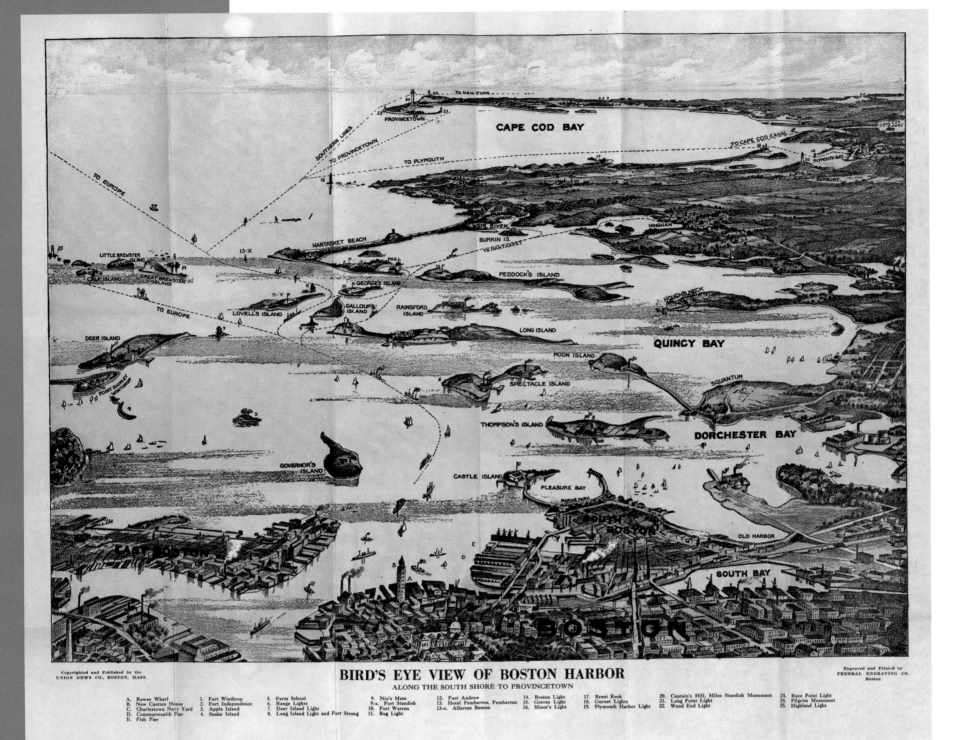

BIRD'S EYE VIEW OF BOSTON HARBOR
ALONG THE SOUTH SHORE TO PROVINCETOWN

A. Rowes Wharf	1. Fort Winthrop	5. Farm School	9. Nix's Mate	12. Fort Andrew	14. Boston Light	17. Brant Rock	20. Captain's Hill, Miles Standish Monument	23. Race Point Light

A. Rowes Wharf
B. New Custom House
C. Charlestown Navy Yard
D. Commonwealth Pier
E. Fish Pier

1. Fort Winthrop
2. Fort Independence
3. Apple Island
4. Snake Island

5. Farm School
6. Range Lights
7. Deer Island Light
8. Long Island Light and Fort Strong

9. Nix's Mate
9-x. Fort Standish
10. Fort Warren
11. Bug Light

12. Fort Andrew
13. Hotel Pemberton, Pemberton
13-x. Allerton Beacon

14. Boston Light
15. Graves Light
16. Minot's Light

17. Brant Rock
18. Gurnet Lights
19. Plymouth Harbor Light

20. Captain's Hill, Miles Standish Monument
21. Long Point Light
22. Wood End Light

23. Race Point Light
24. Pilgrim Monument
25. Highland Light

**A VIEW OF BOSTON HARBOR AND CAPE COD BAY, MASSACHUSETTS, BY THE UNION NEWS CO.,
BOSTON, 1920** Based upon an earlier view by John Murphy, this map shows the
34 Boston Harbour Islands from South Boston to Provincetown, tracing many
of the ferry and steamboat routes. The bays and islands are now within a
national recreation area. With its strategically important location, Castle
Island, just a few miles from the waterfront, is the oldest fortified military site
in British North America (1643). It became joined to the mainland by a
concrete boulevard in the 1930s, and since then dredging has increased the
one-time island's connection to the mainland. **ABOVE**

ROBINSON'S AEROPLANE MAP OF SYDNEY, BY H.E.C. ROBINSON, 1922 Port Jackson is the large, natural bay of Sydney and on 26 January 1788 the British raised their flag at Sydney Cove, where Circular Quay is marked. This aerial view, made less than 150 years later by local map producer H.E.C. Robinson Ltd, shows the area's maritime development and the complexity of transportation links. The map covers the area from Careening Cove in the north to Redfern Park in the south and from Glebe Island in the west to Elizabeth Bay in the east. In addition to the significant public and commercial buildings, named roads and streets, there are parks, ferry routes, places of worship and places of entertainment.

Among Sydney's countless wharves is the world's first iron wharf (built in 1874 for coal loading), the characteristic shape of which meant that the railway lines gently curved on their approach to Darling Harbour station. That southern end of the harbour was filled in during the 1920s, during the same era when work began to build one of Sydney's best-known structures: Sydney Harbour Bridge, which runs from Dawes Point to Milsons Point. Next to the tram sheds on Bennelong Point, to the east of Sydney Cove, is where the city's iconic Opera House now stands. **LEFT**

**DEUTSCHE LUFTHANSA SUMMER FLIGHTPLAN,
1 APRIL–5 OCTOBER 1935** This poster is in the
form of an infographic map for
Germany's national air carrier in Europe
that shows the company's city-to-city
links, the hierarchy of their importance
and the timetable. The map distinguishes,
in black and red, between passenger and
freight services. During the 1930s,
Lufthansa was particularly concerned
with extending the range of its mail
service – it had ventures that gave it
access to cities in China and South
America – and reducing the time taken
for its deliveries. **RIGHT**

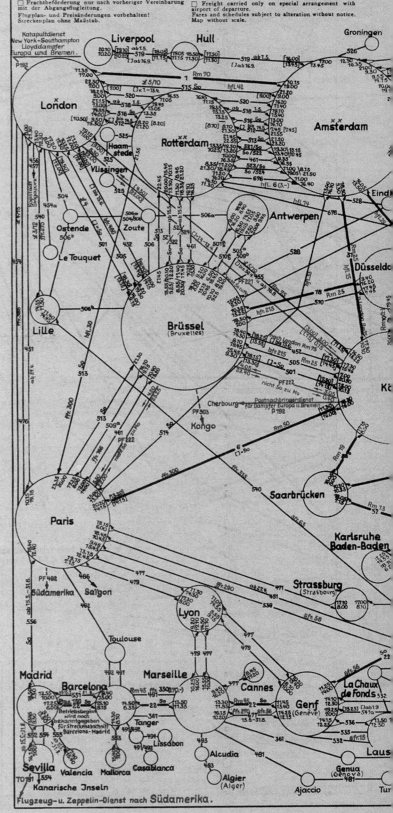

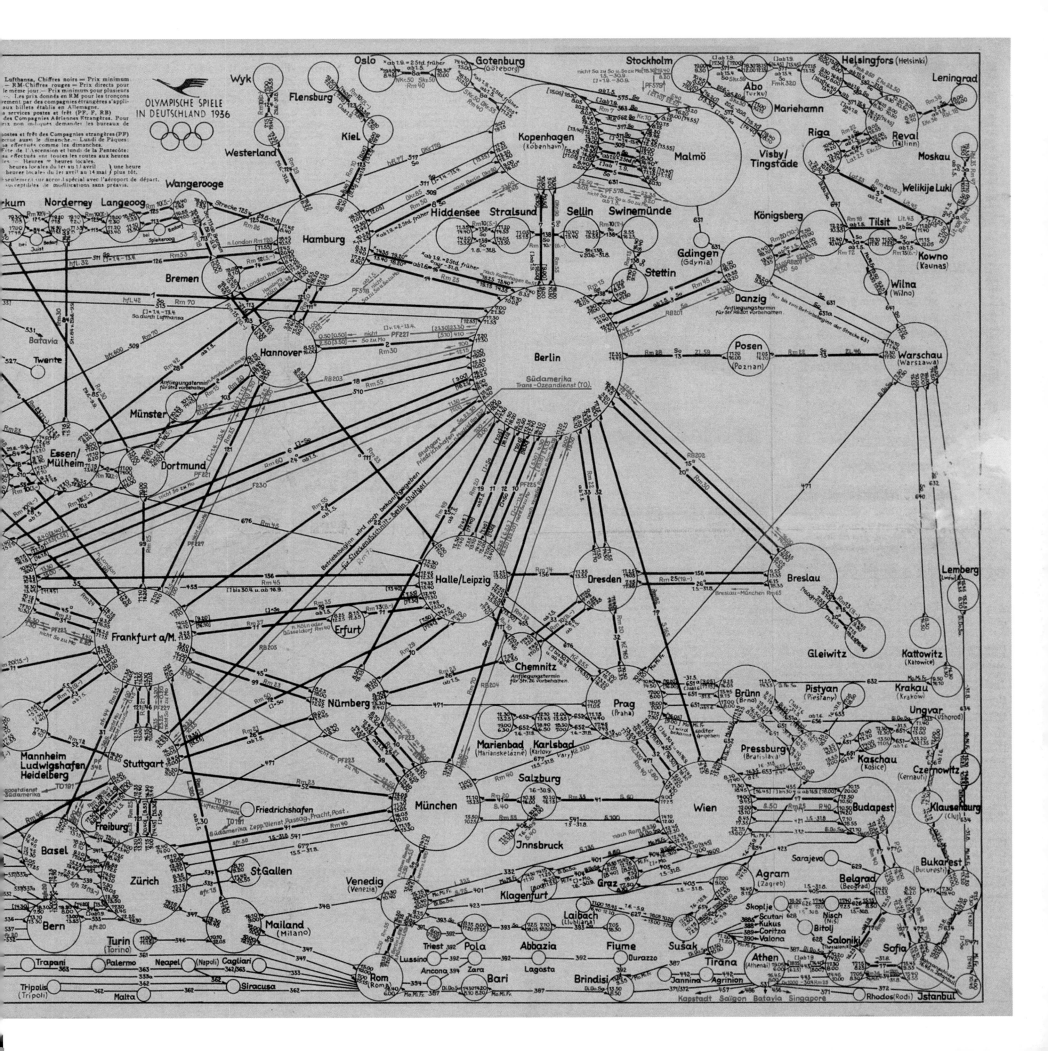

PLAN OF CANBERRA, BY WALTER BURLEY GRIFFIN, 1927 In May 1927 the federal parliament was officially transferred from Melbourne to the new capital of the Commonwealth of Australia, Canberra. The site was chosen in 1909 and in 1912 the competition to design it was won by architect Walter Burley Griffin. Like L'Enfant's plan for Washington, DC, government functions were central features of Griffin's 'ideal city'. He proposed a geometric streetplan radiating from focal points, such as the civic centre, and envisaged an urban setting where the long vistas would enhance the resident's appreciation of the area's natural features. The plan shows how the lakes were expected to be formed from the existing river – a notable feature of Canberra today. But Griffin's relationships with bureaucrats ended in acrimony; by 1920 he was no longer involved and many of his ideas were never implemented. Although Canberra has expanded greatly since, the original plan has not been upheld. **RIGHT**

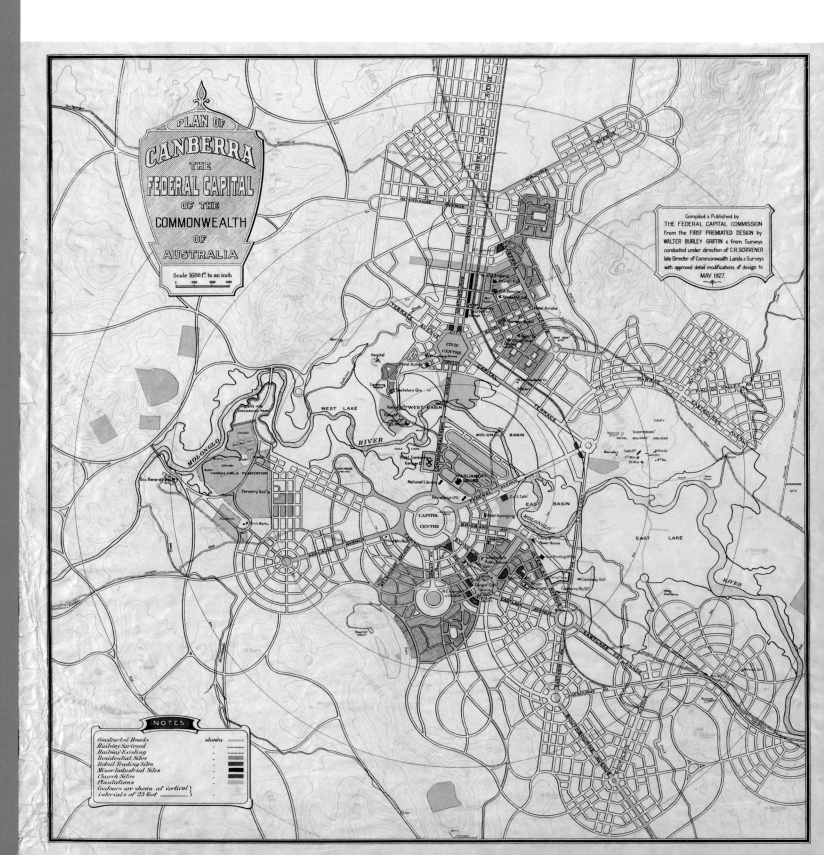

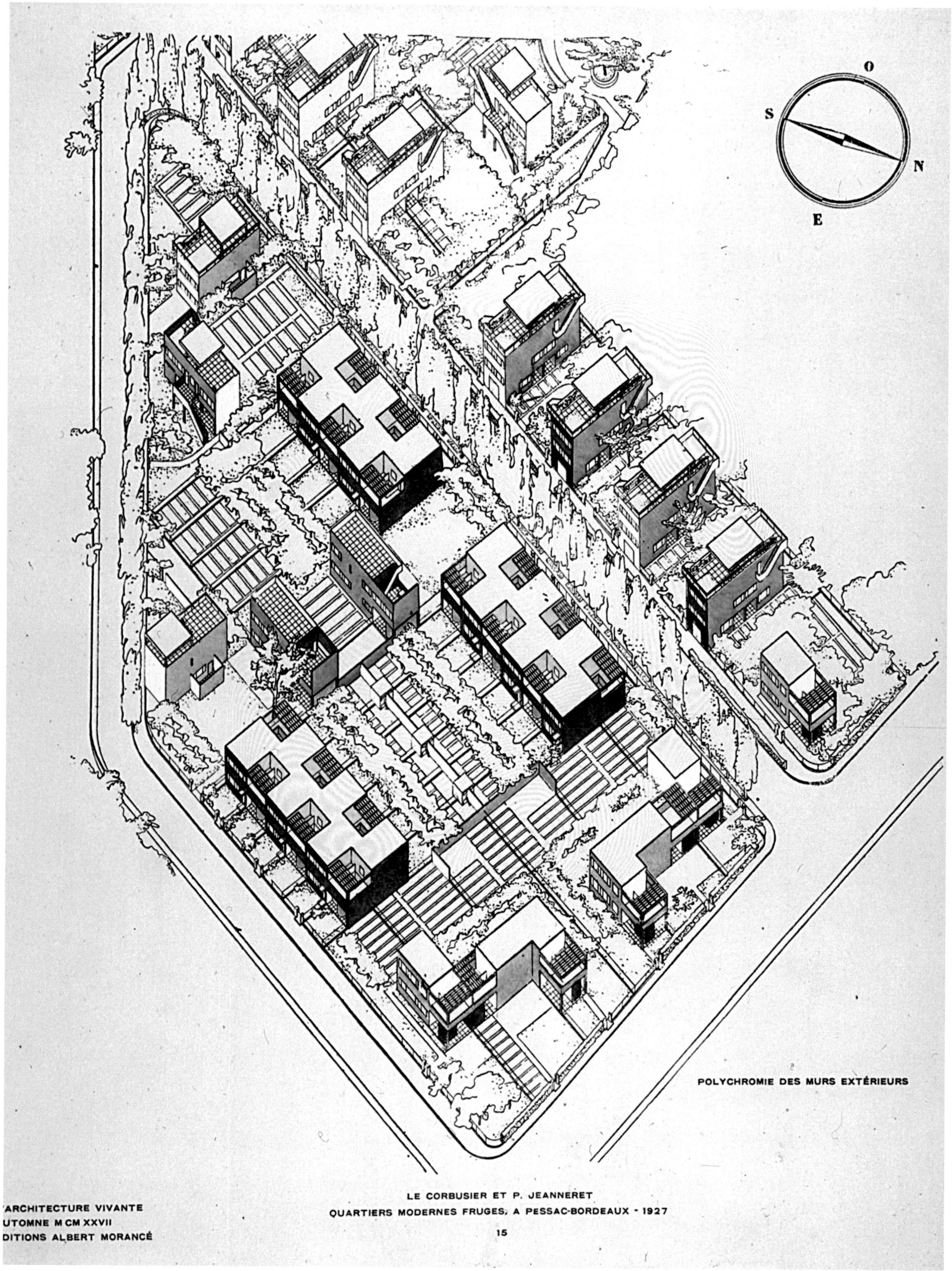

POLYCHROMIE DES MURS EXTÉRIEURS

LE CORBUSIER ET P. JEANNERET.
QUARTIERS MODERNES FRUGES, A PESSAC-BORDEAUX · 1927

ARCHITECTURE VIVANTE
UTOMNE M CM XXVII
DITIONS ALBERT MORANCÉ

15

DESIGN FOR LES QUARTIERS MODERNES FRUGES, PESSAC ESTATE, BORDEAUX, SKETCH FROM *L'ARCHITECTURE VIVANTE* JOURNAL, 1927 This model estate of 51 houses completed in 1926 was sponsored by local industrialist Henry Frugès, who dreamed of somewhere to house his workers that would represent new ideas in tangible form. To realise this utopian vision he chose the rising avant-garde urban architect Charles-Edouard Jeanneret-Gris (better known today as Le Corbusier), whose Modernism, like the Garden City movement, reflected a belief that better buildings and spaces could make better cities. The aim was to combine mass production efficiencies with a module that would satisfy the diverse desires of individuals. Favouring concrete as his chosen material, only the exterior walls were load-bearing so the interior space could be endlessly reconfigured. The architect sought space, light and view, and his cellular designs have proved to be highly adaptable. **LEFT**

Metropolis

SHANGHAI, BY V.V. KOVALSKY, 1935 Produced
for Shanghai Municipal Council and
designed by American journalist and
businessman Carl Crow, this map reflects
his enthusiasm for a cosmopolitan city
where a diversity of nationalities and
cultures, both East and West, mixed –
Chinese, British, American, French,
German, and more. Crow lived in the city
for more than two decades, only leaving
when the Japanese invaded in 1937. He
understood and admired the Chinese,
using his knowledge of local consumers
to develop an advertising agency. He later
wrote an informative account of his time
there, *400 Million Customers*.

The map focuses on the International
Settlement district, but also shows the
French concession and the edge of the
Chinese city below. The Whangpoo
(Huangpu) River is busy with vessels, both
commercial and military, and the main
civic, commercial, religious and leisure
buildings are illustrated and annotated.

The vignettes which border the map
are a mix of Chinese and European
themes, ranging from teahouses, pagodas
and a local funeral procession to the
arrival of missionaries, the building of
tramways and the Hongkong & Shanghai
Bank. **OPPOSITE**

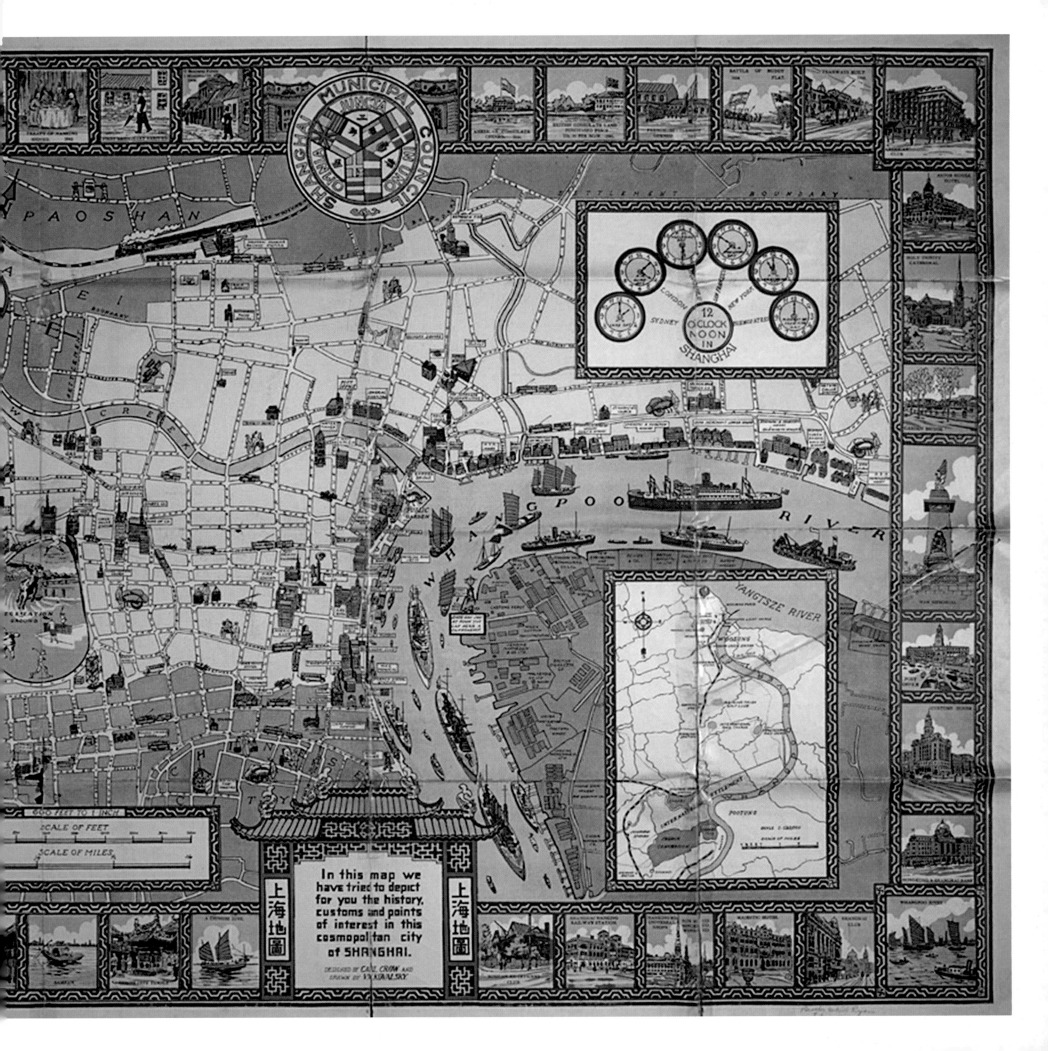

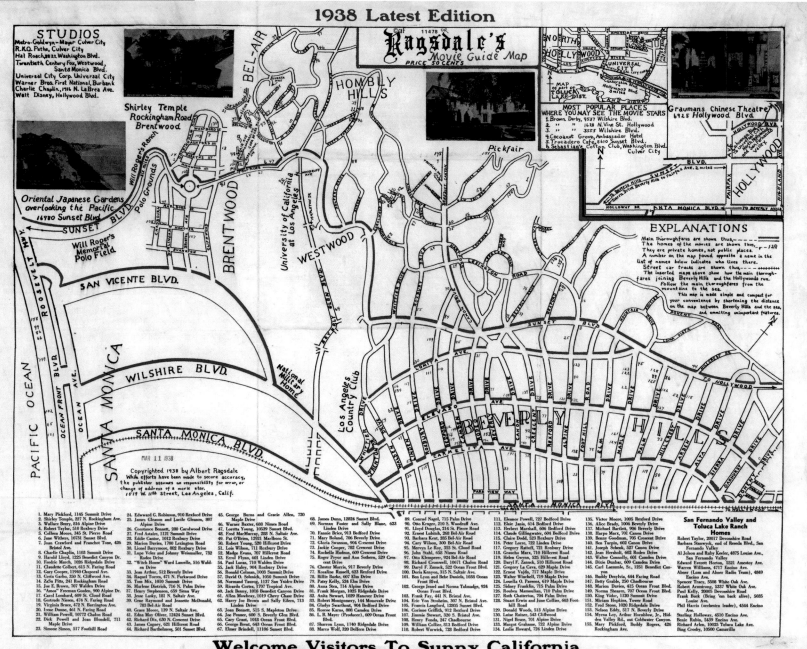

A global era: 1900–2000s

RAGSDALE'S MOVIE GUIDE MAP, 1938 Albert Ragsdale was quick to notice the tourism potential of the emergence of Los Angeles as the entertainment capital of the USA. His hand-drawn, navigable map helped to pioneer what today is a significant element of the city's tourism market: the movie stars and where they live, with an invaluable address listing for more than 150 star names from the glory days of cinema, including Shirley Temple, Joan Crawford, Charlie Chaplin, Greta Garbo, Fred Astaire and James Cagney. **OPPOSITE**

HOLLYWOOD STARLAND, BY DON BOGGS, 1937 Although this 'official moviegraph of the land of stars, where they live, where they work and where they play' is dated to 1937 it depicts the new 'Union Terminal', which opened in 1939. The drawing is far from accurate as a map but it is a great souvenir of a city in its heyday, bordered by good portraits of dozens of famous film stars, from Cary Grant and James Stewart to Bette Davis and Katherine Hepburn, and depicting countless places associated with the movieworld's nouvea-riche. These interesting details include the vast Georgian Revival beach house on the 'Gold Coast' belonging to Marion Davies, William Randolph Hearst's mistress, and reputedly the wildest party venue in town. **LEFT**

BRASILIA: A monument to Modernism

The only 20th-century city to have become a World Heritage site, the new-build federal capital of Brasilia celebrated its fiftieth anniversary in 2010. The purpose of the Modernist creation was to spread Brazil inland from the coast, where Rio de Janeiro lay. The pilot plan for the urban design included a monumental east–west axis as well as a deliberate separation of the civic, administrative space from the residential areas.

BRASILIA, 1960, BASED ON LÚCIO COSTA'S PILOT PLAN
This map detailed the machine-like city of the future's first phase of development, and was printed and distributed by the Shell energy company. Brasilia symbolised the country's mission to modernise, and its location in the interior reflected Brazil's intention to use its natural resources. Demarcated sectors are identified in the key: from banks (8) and industry (23) to multiple residential (29, 30, 32). The superquadras (35, 36) are groups of apartments with schools, retail stores, and open spaces. A peninsula at the northern end of the lake has more upmarket housing (32); a similar area is anticipated on the southern shore. **RIGHT**

BRASILIA DURING THE DRY SEASON, 2001 Brasilia was given the outline form of an aircraft by its urban planner Lúcio Costa in honour of Brazil's role in the early aviation industry, and took just over three years to build (spanning 1956–1960). Located in an area of tropical savannah, an artificial lake, Lago Paranoá, forms part of the city plan. The city's administrative heart lies towards the lake end of the long axis (the aircraft fuselage) at the Plaza of Three Powers, where the executive, legislative and judicial buildings stand. In the north and south (the wings) are residential sections. The lake demarcates the space between the core and the outlying satellite residential districts. Brasilia National Park. **OPPOSITE**

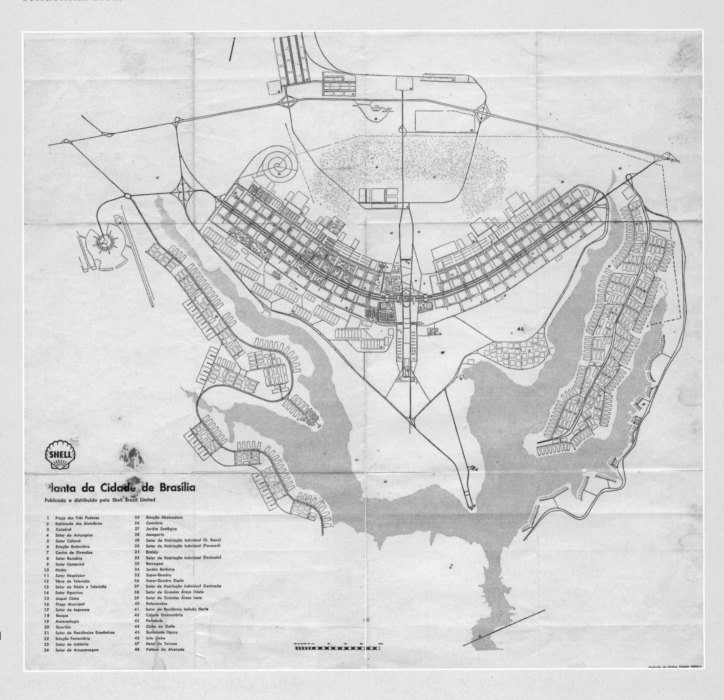

SHELL

Planta da Cidade de Brasília
Publicada e distribuída pela Shell Brazil Limited

1	Praça dos Três Poderes	25	Estação Abaixadora
2	Esplanada dos Ministérios	26	Cemitério
3	Catedral	27	Jardim Zoológico
4	Setor de Autarquias	28	Aeroporto
5	Setor Cultural	29	Setor de Habitação Individual (D. Bosco)
6	Estação Rodoviária	30	Setor de Habitação Individual (Paranoá)
7	Centro de Diversões	31	Ermida
8	Setor Bancário	32	Setor de Habitação Individual (Península)
9	Setor Comercial	33	Barragem
10	Hotéis	34	Jardim Botânico
11	Setor Hospitalar	35	Super-Quadra
12	Tôrre de Televisão	36	Super-Quadra Dupla
13	Setor de Rádio e Televisão	37	Setor de Habitação Individual Geminada
14	Setor Esportivo	38	Setor de Grandes Áreas Oeste
15	Joquei Clube	39	Setor de Grandes Áreas Leste
16	Praça Municipal	40	Embaixadas
17	Setor da Imprensa	41	Setor de Residência Isolada Norte
18	Bosque	42	Cidade Universitária
19	Meteorologia	43	Petrobrás
20	Quartéis	44	Clube de Golfe
21	Setor de Residências Econômicas	45	Sociedade Hípica
22	Estação Ferroviária	46	Iate Clube
23	Setor de Indústria	47	Hotel de Turismo
24	Setor de Armazenagem	48	Palácio do Alvorada

GREATER LOS ANGELES: THE WONDER CITY OF AMERICA, BY K.M. LEUSCHNER, 1932 This pictorial map was published for Los Angeles Metropolitan Surveys, at a time when the city was about to host the Olympic Games (see the rowing venue, bottom right). Much of the USA was in the grip of the Great Depression, but the city, the largest in California and one of the biggest in the country, was still attracting thousands of migrants. Many of the city's most famous buildings and homes, with their distinctive Art Deco styling, were built during this era.

The map does not seem to be aspiring to topographical accuracy (the Los Angeles river is notable by its absence) so much as showing how many places of recreational interest there are in the vicinity of the city, including Redondo Beach, the Hollywood Bowl (1922) and the luxury oceanfront Breaker's Hotel (1926) at Long Beach. The cumulative effect of all these details dedicated to the pursuit of pleasure is to proclaim Los Angeles as the frontier of leisure. Careful scrutiny of the map reveals that Santa Monica had a nude sun baths at its Crystal Pier. RIGHT

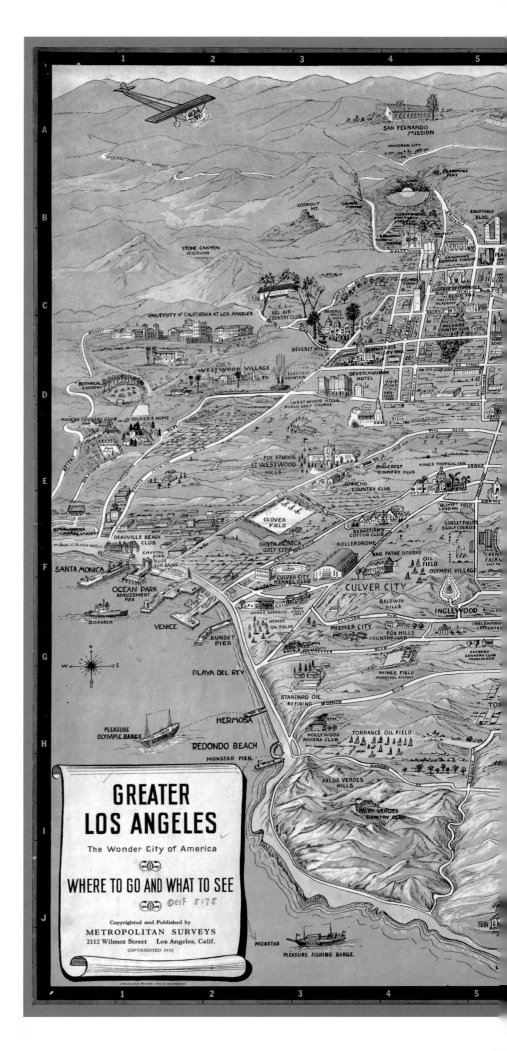

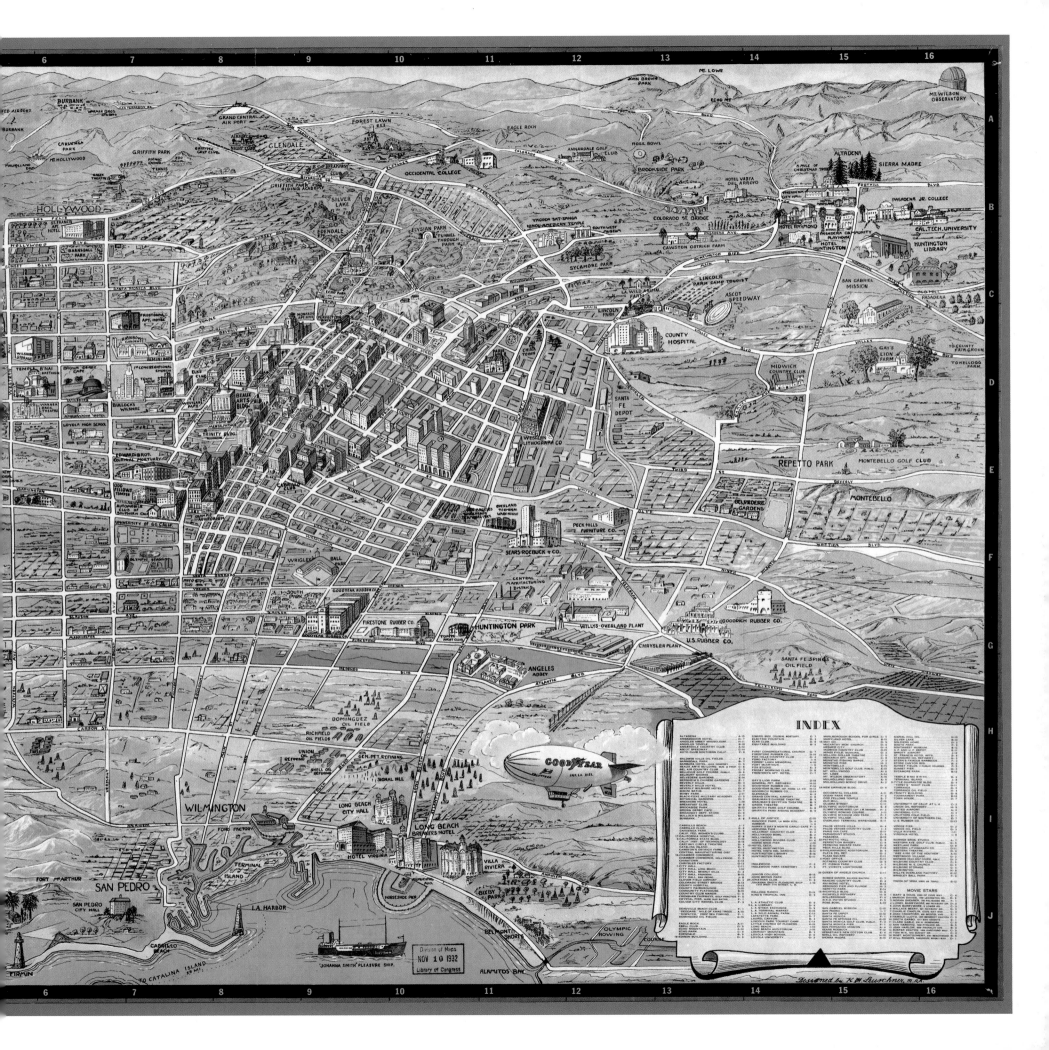

Metropolis

ZONE MAPS OF BERLIN, AIR MINISTRY, UK,

1944–1945 This pair of zone maps, for western and eastern Berlin, was produced by the Air Ministry's Assistant Directorate of Intelligence, specifically the photographic renaissance intelligence unit at RAF Nuneham Park in support of the Allies' strategic air campaign against Nazi Germany. The maps identify Berlin's waterways, woods, open spaces, roads, railways, industrial sites, mines, public utility plants, barracks, and the density of built-up and residential areas. They also highlight the city's administrative districts, major factories, communications and main areas vulnerable to bombing.

One of the most terrible developments of the Second World War was the prolonged use of aerial bombardment to devastate cities. Berlin was the target of more than 350 air raids mounted by British, American and Soviet bombers during the war, which caused massive infrastructure damage and killed perhaps 50,000 people, with many more made homeless. **RIGHT AND OPPOSITE**

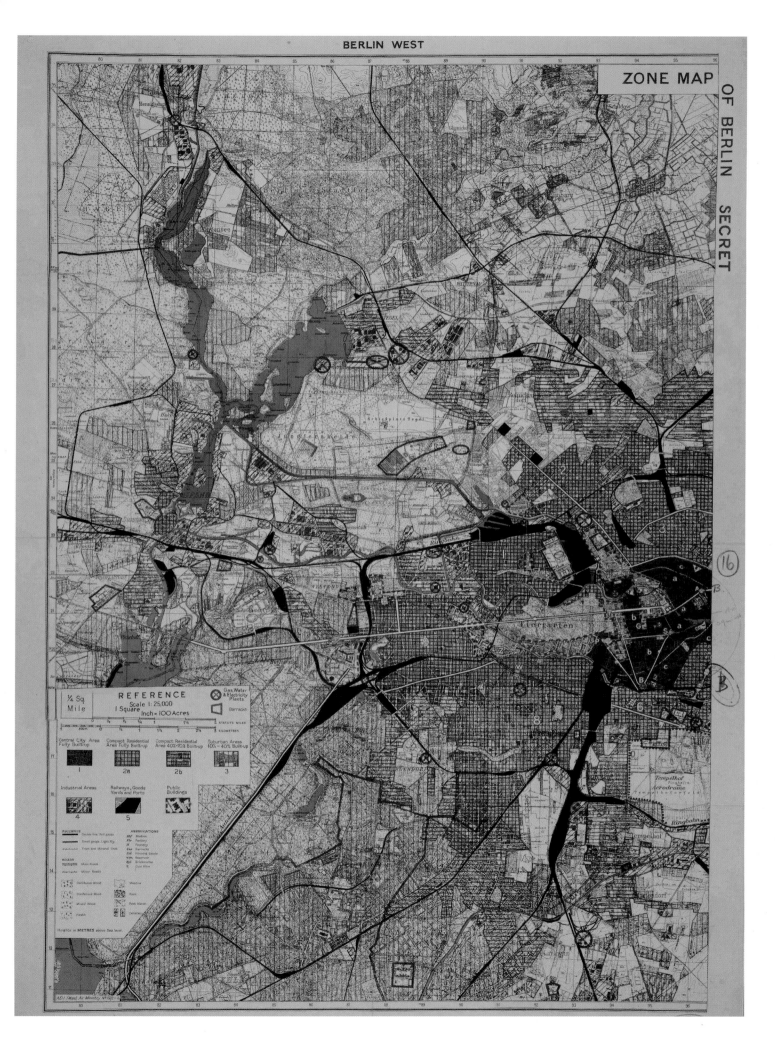

A global era: 1900–2000s

THE ISLAND (DETAIL), BY STEPHEN WALTER, 2008
This 'map' shows London, or rather a detail from a much larger map of the city titled *The Island*, which presents the city as an intensely personal place filled with often hidden meaning, which reflects what the artist, Londoner Stephen Walter, deems to be interesting or mundane. In some respects *The Island* is a form of pictorial map, but rather than the grand bird's-eye perspective the viewer is drawn in by the microscopic detail, which will mean most to Londoners themselves, who will be better able to pick up the countless references.

The artist has included a mass of local and autobiographical information in words and symbols, often making satirical comments about locations. For example, Buckingham Palace is a crown, annotated 'one of the homes of the expensive family'. **RIGHT**

DATA DRIVEN DETROIT, 2010 Detroit, Michigan, is one of the starkest examples of the rise, fall and possible rebirth of a city. Halved in population since its heyday as the Motor City, between 2000 and 2010 the depopulation rate of Detroit was 25 per cent, resulting in vacant properties prone to decay and arson. However, as these two maps reveal it can be difficult to identify correlations within the urban fabric – for example, between fire damage and depopulation. Many of the city's neighbourhoods continued to thrive despite a flight to the suburbs. For example, centrally located historic Woodbridge, to the southeast of where highways 96 and 94 intersect, contains good Victorian housing that has meant that gentrification has begun. Modern technology and the wealth of data available mean that Detroit, the poorest big city in the USA, has generated a wealth of thematically mapped information. **RIGHT AND OPPOSITE**

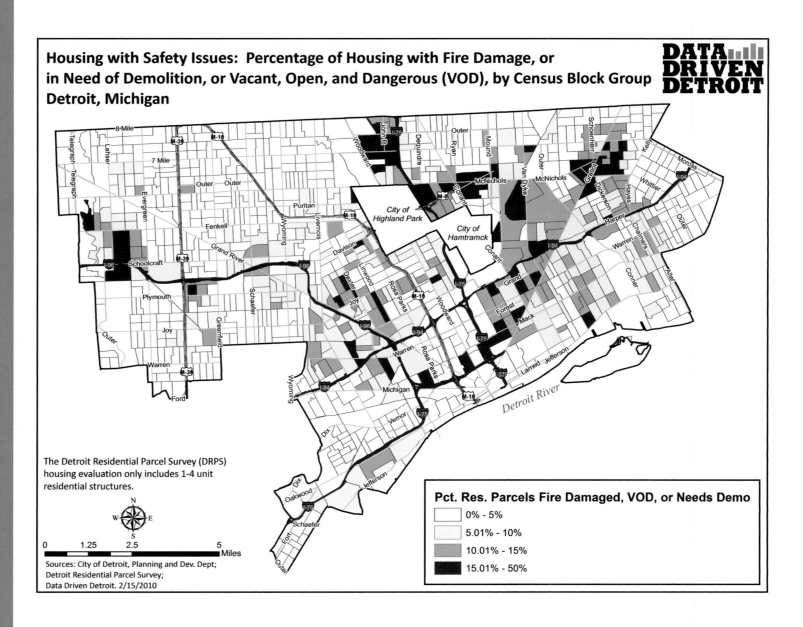

Housing with Safety Issues: Percentage of Housing with Fire Damage, or in Need of Demolition, or Vacant, Open, and Dangerous (VOD), by Census Block Group Detroit, Michigan

The Detroit Residential Parcel Survey (DRPS) housing evaluation only includes 1-4 unit residential structures.

0   1.25   2.5   5
                          Miles

Sources: City of Detroit, Planning and Dev. Dept;
Detroit Residential Parcel Survey;
Data Driven Detroit. 2/15/2010

Pct. Res. Parcels Fire Damaged, VOD, or Needs Demo

0% - 5%
5.01% - 10%
10.01% - 15%
15.01% - 50%

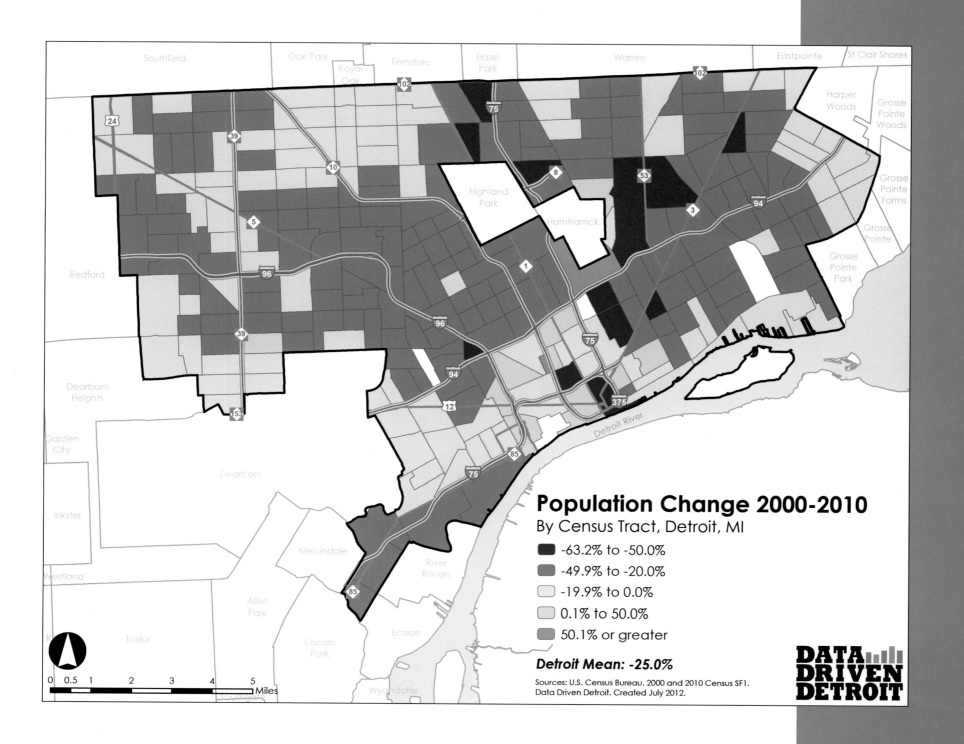

Population Change 2000-2010

By Census Tract, Detroit, MI

- ■ -63.2% to -50.0%
- ■ -49.9% to -20.0%
- ▫ -19.9% to 0.0%
- ▫ 0.1% to 50.0%
- ▪ 50.1% or greater

Detroit Mean: -25.0%

Sources: U.S. Census Bureau, 2000 and 2010 Census SF1.
Data Driven Detroit. Created July 2012.

DATA DRIVEN DETROIT

0 0.5 1   2   3   4   5 Miles

Shanghai is arguably the most stunning example of the speed with which a city can be turned from a regional centre into a bustling, international megacity. As such Shanghai represents the city as the agent of modernisation and change. In the 1930s New York had more skyscrapers than the rest of the world combined, with about 200. Since 1990 more than double that number of skyscrapers has been built in the historic part of Shanghai, and the city is believed to have more than 1000 buildings with more than 30 storeys. The 1:500 scale model at the Shanghai Urban Planning Exhibition Centre envisages the city in 2020. Such population growth and urban development has been accompanied by massive demolition and the forced resettlement of hundreds of thousands of people, but the model emphasises the view of the bigger picture: a Chinese city with a history of international commerce has its eye set firmly on the future. PREVIOUS PAGES

The city of the future has always been a theme in urban mapping. Indeed, Heaven as the City of God was a subliminal religious message worldwide, most prominently Christianity with St Augustine's fundamental work *The City of God* (c.412–27).

During the last half a millennium, secular visions of the future became increasingly more important in thought about cities. They drew on the practical issue of planning and on notions of perfectibility, if not utopianism, that seemed to offer a future in terms that could be understood. In this process, mapping overlapped with both governments' desire for documentation and imaginative literature, as with Louis Sébastien Mercier's utopian novel *L'An 2440* (1770), in which the urban space ranged to include adequate grain stockpiling and a monument – in the Paris of the future - depicting a black man, his arms extended, a proud look in his eye, surrounded by the pieces of twenty broken sceptres, while atop a pedestal reading '*Au vengeur du nouveau monde*' ('avenger in the new world').

AUTHORITARIAN VISIONS

Historically, the feature of planning under established authoritarian regimes was that present-day power and ideology was inscribed onto the townscape of the future. This happened, for example, with Christianity in Baroque Rome, and also under radical counterparts such as the French revolutionaries of the 1790s, as well as their Soviet, Nazi and Chinese communist counterparts in the 20th century. Albert Speer planned to honour Hitler and provide the anticipated thousand-year Reich with a gleaming new capital by transforming Berlin into Germania. Benito Mussolini ploughed a major thoroughfare through the ancient neighbourhoods of Rome; he also planned a monumental town as a gateway to an expanded Rome.

Work began in 1938 on E42, or the Esposizione Universale di Roma, intended to be ready for a world fair in 1942 that would celebrate 20 years of Mussolini's rule. These buildings were completed in the 1950s and survive today as Rome's EUR district. Similarly, the capture of Madrid in 1939 by Nationalist forces under Franco at the close of the Spanish Civil War was followed by schemes to transform the city, but in the wake of the war these plans fell victim to a lack of money. Nicolae Ceauçescu, the Romanian dictator from 1965 to 1989, rebuilt large parts of Bucharest, and Beijing has seen wholesale redevelopment under China's Communist government.

EXEMPLARY CITYSCAPES

Generally within more benign political contexts, a similar process can be seen with the creation of new cities, notably new capitals – as in the case of Madrid in 1561. This process of foundation became more common as empires were replaced by independence, although other factors also led to the establishment of new capitals. These included Washington, DC, USA; Canberra, Australia; Brasilia, Brazil; and Abuja, Nigeria. Chandigarh, the new capital of the Punjab, was designed by Le Corbusier. In each case, the plan determined to produce an exemplary cityscape, but the reality was often very different. Brasilia was located in the inland centre of Brazil to symbolise national unity and escape from the politics of Rio de Janeiro and São Paulo. The new capital was developed in a functional fashion by Oscar Niemeyer, a protégé of Le Corbusier, but the city is now surrounded by shanty towns, the population of which provide the workforce Brasilia needs to operate.

The quest for an exemplary built environment motivated the New Town movement in 20th-century Britain and the arguments about the provision of

FROM THE *CITY OF GOD* BY AUGUSTINE OF HIPPO, C.426 After the sack of Rome by the Visigoths in 410, many citizens felt that their lapsed faith had caused the event. Augustine argued that history was a conflict between the Earthly City and the City of God, the mystical, heavenly New Jerusalem – a model form of urban life (depicted at the top of the painting) into which people could ascend, depending on whether they adhered to the seven Christian virtues or succumbed to temptation from the seven deadly sins. The illustration is from a translation by Raoul de Presles, c.1469-73 **LEFT**

Metropolis

'KING'S DREAM OF NEW YORK',
BY MOSES KING, 1908 In the early 20th
century, journalists were forever
prophesying New York's future – after
all, the greatest city in the USA had seen
quite remarkable changes since 1808. In
1908 the Singer Building (depicted rear
left) was the world's tallest, but in this
skyscraper vision of the urban world to
come it has been overshadowed by taller
buildings and elevated railroads connect
the buildings and planes fill the skies.
RIGHT

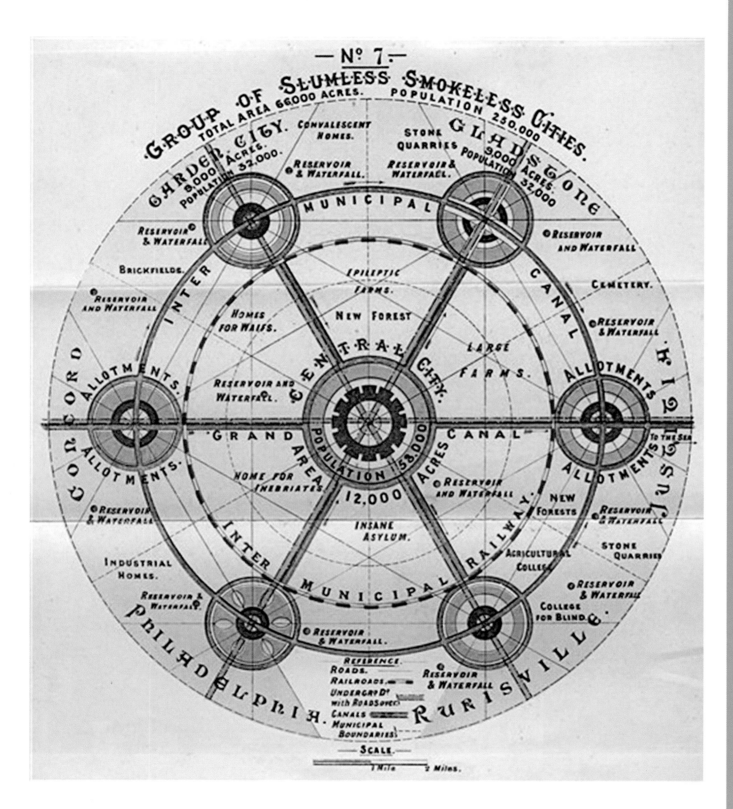

SOCIAL CITY, BY EBENEZER HOWARD, 1902 This diagram is taken from his treatise on *Garden Cities of To-morrow*, Howard argued for the cooperative development of well-planned concentric towns of limited size and protectively surrounded by a permanent belt of agricultural land. His radial diagram – it was not so much a precise plan as a concept to inspire one – shows how green spaces (countryside, parks, gardens, allotments) might be maintained alongside residential areas and for there to be means of 'rapid communication' between places. He thought that by setting a limit to size, new towns could be added in a form of connected cluster. In 21st century Britain, the garden city movement's vision of society and nature in union is once again informing discussions of new urban development. **LEFT**

public housing. Indeed, it was believed that appropriate urban provisions would ensure not only the means to meet the physical and material demands of the population but also that lifestyles developed that were in keeping with being moral citizens. The rapid spread of cities in accordance with government planning then became a characteristic feature of much of the world in the 20th century.

One of the solutions to meeting the growing need for housing – the cheap, mass-production construction of urban high-rises, which required only modest amounts of land – is difficult to appreciate in traditional cartography, with its emphasis on mapping from directly overhead.

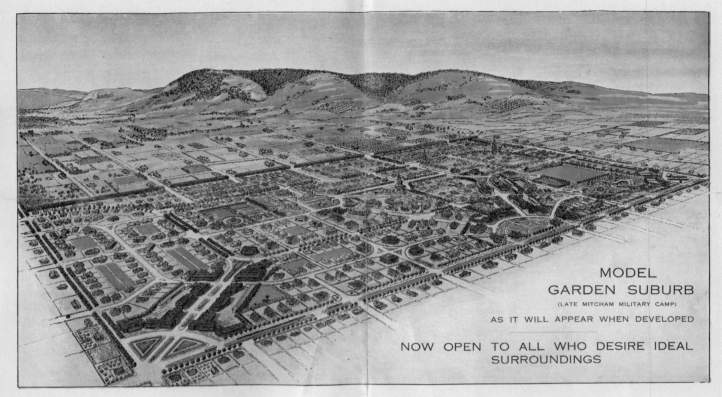

SOUTH AUSTRALIAN GOVERNMENT
COLONEL LIGHT GARDENS

MODEL
GARDEN SUBURB
(LATE MITCHAM MILITARY CAMP)
AS IT WILL APPEAR WHEN DEVELOPED

NOW OPEN TO ALL WHO DESIRE IDEAL
SURROUNDINGS

COLONEL LIGHT GARDENS SALES BROCHURE, MITCHAM, GREATER ADELAIDE, 1921 This South Australia suburb was named after the first surveyor of the state, Colonel William Light, who chose the site for Adelaide and began his survey work there in 1836–1837, before he died of tuberculosis in 1839.

In 1914 South Australian Charles Reade advocated a model garden suburb with land use zoning in the state and by 1921 a plan had been approved for a scheme under the name Colonel Light Gardens. By 1927 more than 900 homes had been built, but the scale of the politically inspired revised development was greater than that originally envisaged. After 1927 a further area was added that became known as Reade Park. RIGHT

How best to ensure housing quality and social cohesion remain important themes today, while city planning is arguably more important than ever as an issue because the proportion of the world's rising population who live in cities continues to increase. Indeed, the United Nations suggests that while more than half of the world's population of seven billion live in cities in 2012, by 2030 the number will be more than five billion. There are projections that three-quarters of the global population by 2050 will be living in cities, with most of the increase occurring in Asia and Africa.

Indeed, the 20 biggest cities in 2015 in terms of urban agglomerations, a criteria that leads to variations in the data based on differing definitions, were: Tokyo–Yokohama, Jakarta, Delhi, Manila, Seoul–Incheon, Shanghai, Karachi, Beijing, New York, Guangzhou–Foshan, São Paulo, Mexico City, Mumbai, Osaka–Kobe–Kyoto, Moscow, Dhaka, Cairo, Los Angeles, Bangkok and Kolkata (Calcutta). Moscow, estimated at 16 million, is the only European city in the list. Of the thousand largest urban areas, around 56 per cent is in Asia. The world now has 34 megacities (population of 10 million-plus). The need for new cities – China requires hundreds to cope with migration from the land – is a major factor driving research into more and better sustainable forms of urban development.

Mapping will have to adapt to the challenges presented by this pace of change, including being fully aware of the relationship between new technologies and developing urban needs. For example, in 2012 Google Inc. built a fibre-optic broadband network that was able to deliver the fastest Internet speeds in the USA, and the company began to offer the service in Kansas City. Google divided the city into 204 districts, or fibrehoods, and asked those who wanted the service to register in advance. Google agreed to provide the service to the 46 fibrehoods with the greatest concentration of consumers willing to pay. Thus, Kansas City can be mapped on a new basis.

May Live to See

May Solve Congestion Problems

LIVING QUARTERS AND PLAYGROUNDS

AIRCRAFT LANDING FIELDS

SCHOOLS

OFFICES

LEVEL FOR PEDESTRIANS

RESTAURANTS

SLOW MOTOR TRAFFIC

RAMP

FAST MOTOR TRAFFIC

GARAGES

ELECTRIC TRAINS

SPIRAL ESCALATORS

FREIGHT TUBES

How You May Live and Travel in the City of 1950

Future city streets, says Mr. Corbett, will be in four levels: The top level for pedestrians; the next lower level for slow motor traffic; the next for fast motor traffic, and the lowest for electric trains. Great blocks of terraced skyscrapers half a mile high will house offices, schools, homes, and playgrounds in successive levels, while the roofs will be aircraft landing-fields, according to the architect's plan

'THE WONDER CITY YOU MAY LIVE TO SEE', BY HARVEY W. CORBETT, 1925 Looking just a generation ahead (1950), this section of architect Harvey W. Corbett's artwork from the August 1925 edition of *Popular Science Monthly* depicts 'aircraft landing fields' on skyscraper rooftops and four separate street levels for pedestrians, slow and fast moving motor traffic, and subway electric trains accessed via spiral escalators. Freight moves through tubes. Anticipating more and more people congesting the urban space, the separation of people and motorised transportation was the thinking of the day, although the actual development of cities turned out rather differently. LEFT

Metropolis

METROPOLIS, BY FRITZ LANG, 1926 The
towering architectural–industrial city to
come, as imagined by Fritz Lang in his
adaptation of screenwriter Thea von
Harbou's fantasy of man and machine in
the future. By the 1920s the idea of a
vertical city had become widespread in
American popular culture. Fritz Lang later
claimed that it was the first sight, in 1924,
of the towering metropolis of the
Manhattan skyline at night – with its
'glaring lights and the tall buildings' –
from aboard SS *Deutschland* in the
harbour that provided him with his
inspiration for the look of the film. RIGHT

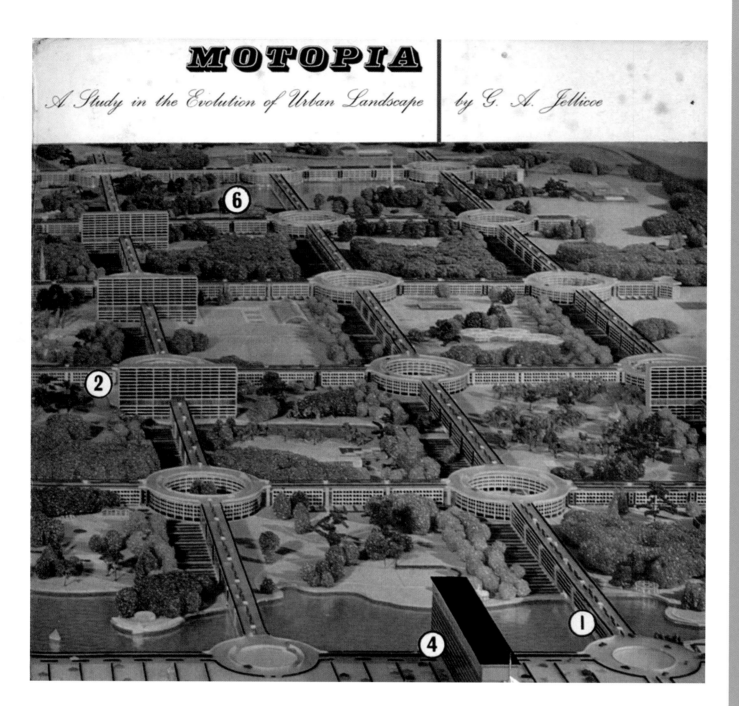

MOTOPIA
A Study in the Evolution of Urban Landscape | by G. A. Jellicoe

MOTOPIA: A STUDY IN THE EVOLUTION OF URBAN LANDSCAPE, BY GEOFFREY ALAN JELLICOE, 1961 In 1959, at around the same time as Oscar Niemeyer was designing buildings within Lúcio Costa's Brasilia, British architect Jellicoe had a vision for a completely new type of layered town in which cars would be in the air. This was to be a dormitory town with no heavy industry, built to the west of London from the ground up. Cars would move about on elevated roads that connected the buildings within a grid layout. Each roof had a roundabout. Pedestrians would have moving walkways and ground level would effectively be a series of parks. Jellicoe proclaimed that he was 'separating the biological elements from the mechanical'. **LEFT**

MIND MAPS AND PLANNING MAPS

Cities serve to highlight the issue of the need to managing unprecedented growth, which is now a key element in the global future – not least because of concerns about the social and political consequences of unsuccessful expansion. Indeed, dystopias are now very much seen in terms of urban chaos and an unmanaged cityscape – a theme that links imaginative works, such as the vision of Gotham City (New York) in the *Batman* films, to more grounded studies. The city, especially the inner city, dominates much fiction, and it is in urban locations that change affects human

drama with the most intensity and pungency. The two maps that are therefore of particular issue are: first, the mental maps that represent fears of dystopias, notably of inner cities; and, second, the maps of city planners. The latter also indicate the extent to which the living city is always a work in progress.

Alongside significant shifts in the urban world, there have come major changes in the technology of mapping, notably as a result of digitalisation. Computer technology has transformed and given greater flexibility to the acquisition, storage and use of cartographic information. Before, maps were printed

**QUEEN'S MUSEUM PANORAMA OF NEW YORK,
1964** Conceived by Robert Moses, New
York's master planner, and built originally
as a scale model (1:1200) exhibit piece for
the 1964 World's Fair, the Panorama took
100 craftsmen three years to make and
assemble its 830,000 wood and plastic
buildings. Covering all five boroughs, the
mission was to make the representation
as accurate as possible so that it could
then be used as a planning tool, enabling
an unrivalled overview of urban contexts
and neighbourhoods. RIGHT

HAMBURG'S GREEN NETWORK PLAN, 2013.
Hamburg, Germany's second-biggest city,
is planning a Green Network that will
cover about 40 per cent of the city's area.
Being able to enjoy nature where you live
is one of its aims. The plan identifies
several types of green space: open space
(woods and agricultural land) in light
green; space being put to a specific use
(allotments and cemeteries, for example)
in medium green; and parks (borough or
district parks) in dark green. Vertical
green-on-white lines are green residential
space; horizontal lines mean urban
recreational areas. Within this citywide
area more cyclists and pedestrians will be
encouraged. OPPOSITE

from a master printing copy that was made up of
several flaps or overlays, each containing part of the
relevant whole. In contrast, computers store material
previously divided between several flaps in a database,
and maps are produced as a raster that consists of many
minute dots. These record symbols as dots, lines or
surfaces, and the information can be presented on
screen, where it can be edited, and the entire map
printed out in one process. Geographic base maps are
stored as digital images and the separation of storage
(the digital map) and display makes the analysis and
presentation of the data more flexible.

Digitisation therefore makes the mapping of cities
(and other topics) far easier. All the elements can be
readily modified. Updating is far easier, and it is simple to
modify bases. Different projections, perspectives, sizes
and centrepoints can readily be used. The geo-coding of
digital databases and the digitisation of the Earth are
crucial, making it possible to combine data files,
cartographic coordinate files and statistical mapping
software in an automated statistical mapping system.

Digital systems are being fed by more data, with
Google Earth providing detailed three-dimensional
imagery of the world's major cities. This capability
leads to fears of surveillance and control, as well as to
the possibility of individuals engineering 'mashup'
maps that blend Google base maps with other data. In
all cases, the ambiguity, centrality and diversity of
mapping is emphasised while the city increasingly
becomes the focus of activity.

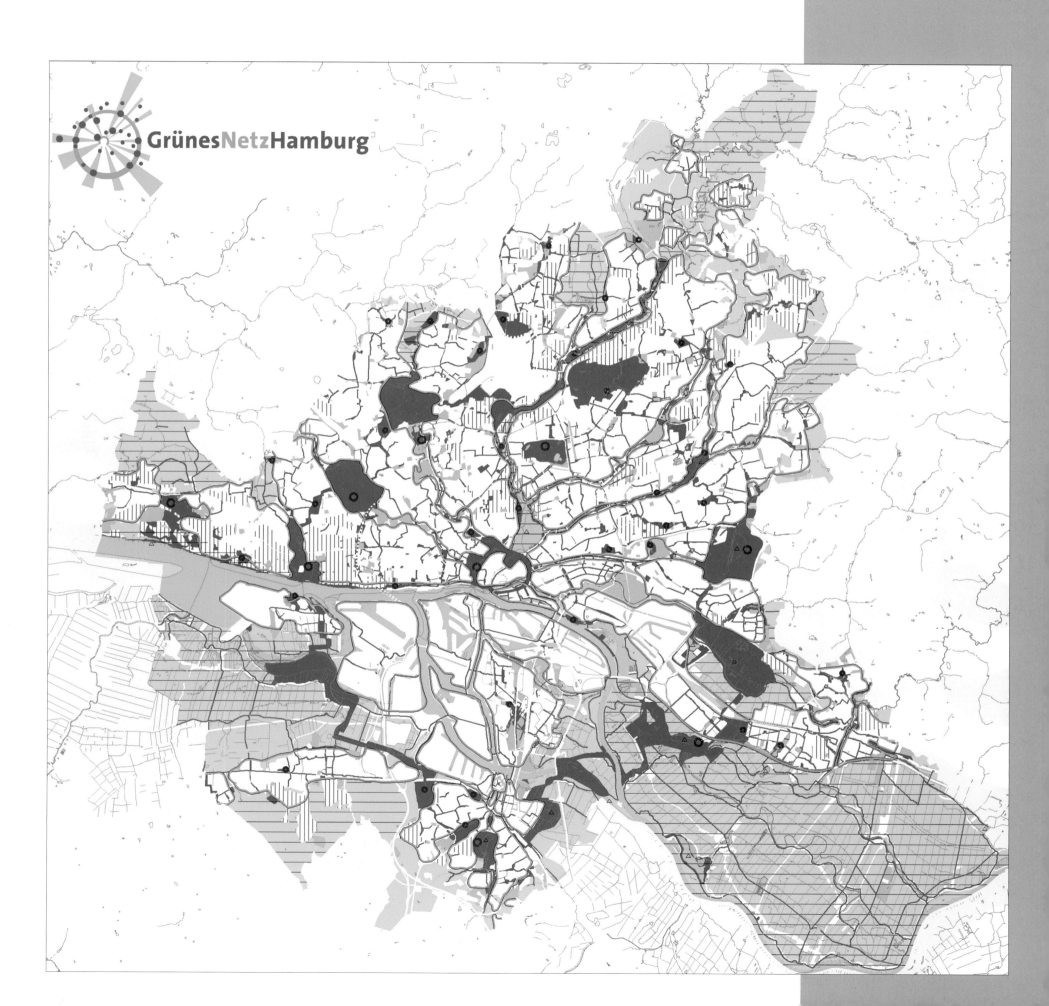

GrünesNetzHamburg

ECO-CITIES: Green replaces grime

The need for more sustainable living and the invention of technologies – such as solar- and wind-energy harvesting and 'smart' sensors that lower consumption of electricity and water – that make this high-function, low-footprint existence possible has resulted in several environmentally friendly new-build urban dreamscapes in recent years. Such mini-utopias have their critics, but the world's older cities may yet be the beneficiaries.

A MODEL OF DONGTAN, CHONG MING ISLAND, NEAR SHANGHAI, 2005 Migration from rural areas means China needs 400 cities by 2020 for 300 million people. Dongtan was meant to be the sustainable, energy-efficient, integrated urban development China needs, but the grand vision set out in 2005 has yet to materialise. **BELOW**

AN ARTIST'S IMPRESSION OF MASDAR, ABU DHABI, 2010 Claiming to be the world's first carbon-neutral city, Masdar is a futuristic showcase in the desert for an oil-rich state committed to research into renewable energy, sustainability, and clean-energy technology. **RIGHT**

SONGDO INTERNATIONAL BUSINESS DISTRICT, SOUTH KOREA, 2009
Songdo is built on a tidal flats landfill site 30 miles from congested Seoul as an Asian smart hub with all the things people want from city life, from good schools and open spaces to ease of movement or walkability and culture. Linked by bridge to Incheon international airport, the city has six core sustainable design goals: open and green space; transportation; water consumption and re-use; carbon emissions and energy use; material flows and recycling; and sustainable city operations. **LEFT AND BELOW**

LIST OF MAPS

INDEX

IMAGE CREDITS

2 Library of Congress; 5 R. Merlo/De Agostini/Getty Images; 7 Universal History Archive/UIG via Getty images; 10 Werner Forman, Universal Images Group/Getty; 13 Dea Picture Library. De Agostini/Getty; 14 The Art Archive/Alamy; 15 De Agostini Picture Library/Getty; 16 www.BibleLandPictures.com/Alamy; 17 Gianni Dagli Ortis/Corbis ; 18 PBL Collection/Alamy; 20 The Art Archive/Alamy; 22 Bodleian Library; 23 De Agostini Picture Library/Getty; 24 Italian School, Getty; 26 Werner Forman/Universal Images Group/Getty; 28 Gianni Dagli Ortis/Corbis ; 29 Newberry Library Chicago, Bridgeman; 30 Alinari Archives, Florence/Mary Evans Picture Library; 32 Historic Cities Research Project © Ozgur Tufekci; 35 Private Collection/The Stapleton Collection/Bridgeman Images; 36 British Library; 37 The Art Archive/Alamy; 38 Library of Congress/Jay I. Kislak Collection; 39 The National Archives; 40 Lifestyle pictures/Alamy; 42 British Library; 43 Alinari Archives/Corbis; 44 Museo de Firenze Com'era, Florence, Italy/Bridgeman Images; 46 Mary Evans/Interfoto; 48t Private Collection/Bridgeman; 48b D'Agostini/R.Merlo/; 49 Images and Stories/Alamy; 50 The Stapleton Collection/Bridgeman; 51 Islamicarts.org; 52 The Stapleton Collection/Bridgeman; 53 Wien Museum Karlsplatz, Vienna, Austria/Ali Meyer/Bridgeman; 54 D'Agostini/Getty; 55t D'Agostini/Getty; 55b Historical Picture Archive/Corbis; 56 De Agostini Picture Library/Bridgeman; 57 Bettman/Corbis; 58 De Agostini/Getty; 61 De Agostini/Getty; 62 De Agostini/Getty; 63 De Agostini/Getty; 64 AKG Images/Historic Maps; 65 AKG Images/Universal Images Group Archive; 66 New York Public Library; 67 PBL Collection/Alamy; 68 Hulton Archive/Getty; 68 Heritage Images/Hulton Archive/Getty; 69 Hulton Archive/Getty; 70 Mary Evans/Iberfoto; 71 Classic Image/Alamy; 72 Heritage Image Partnership/Alamy; 74 AKG/British Library; 75 AKG/Historic Maps; 75 AKG/Historic Maps; 76 Brooklyn Museum of Art, USA/Gift of W. W. Hoffmam/Bridgeman; 77 SSPL/Getty; 78 SSPL/Getty; 79 Library of Congress; 80 British Library/Robana; 82 NDF/Alamy; 85 AKG Images; 86 Buyenlarge Archive Pictures/Getty; 88 British Library; 90 British Library; 91 Hulton Archive/Getty; 92 British Library; 93 British Library; 93 Biblioteque Nationale de France; 94 Library of Congress; 96 Paul John Fearn/Alamy; 97 AKG Images/Historic Maps; 97 Library of Congress; 98 British Library; 100 Library of Congress; 102 De Agostini/Getty; 103 National Library of Scotland; 104 Corbis; 105 Library of Congress; 106 Wikipedia Commons; 108 AKG Images/Historic Maps; 110 Library of Congress; 111 AKG Images/Historic Maps; 112 British Library; 115 D'Agostini/Getty; 116 University of British Columbia, Beans Collection; 117 SSPL/Getty; 118 Library of Congress; 120 Library of Congress; 121 Library of Congress; 122 British Library; 123 British Library; 124 Austrian Archives/Corbis; 125 Library of Congress; 126 Mary Evans Picture Library/Mapseeker Publishing; 128t Library of Congress; 128b Library of Congress; 129 Library of Congress; 130t University of Glasgow Libraries; 130b British Library; 131 British Library; 132 Wikipedia Commons; 134 Library of Congress; 135 Getty/Historic Map Works; 136 Library of Congress; 138 Library of Congress; 139 Library of Congress; 140 Library of Congress; 141 Library of Congress; 142 Library of Congress; 142 British Library; 143 British Library; 144 Library of Congress; 146 Library of Congress; 147 Private Collection/Archives Charmet/Bridgeman Images; 148 Library of Congress; 149 Library of Congress; 149 Library of Congress; 150 Library of Congress; 151 Library of Congress; 152 Library of Congress; 154 Library of Congress; 154 Library of Congress; 155 Library of Congress; 155 Library of Congress; 156 Library of Congress; 157 Library of Congress; 158 Sainsbury Instit Cortazzi; 159 University of British Columbia, Beans Collection; 160 Library of Congress; 161 Library of Congress; 164 Library of Congress; 165 Library of Congress; 169 Public domain; 170 Chicago History Museum/Getty Images; 171 Chicago History Museum/Getty Images; 172 Library of Congress; 173 AP/Press Association Images; 174 Library of Congress; 175 Library of Congress; 176 Library of Congress; 178 The Granger Collection NYC/Topfoto; 181 The National Archives/HIP/Topfoto; 182 Claro Cortes IV/Reuters/Corbis; 183 Sebastian Kahnert/dpa/Corbis; 184 Library of Congress; 185 Library of Congress; 186 The Royal Aeronautical Society (National Aerospace Library)/Mary Evans; 188 Library of Congress; 189 Bibliotheque des Arts Decoratifs, Paris, France/Archives Charmet/Bridgeman Images; 190 Boston Public Library; 192 Library of Congress; 193 Library of Congress; 194 Library of Congress; 195 © NASA/Corbis; 196 Library of Congress; 198 The National Archives; 199 The National Archives; 200 Stephen Walter; 202 Data Driven Detroit; 203 Data Driven Detroit; 204 Christian Kober/Robert Harding World Imagery/Corbis; 207 Bibliotheque Nationale, Paris, France/Archives Charmet/Bridgeman Images; 208 Mary Evans Picture Library; 209 Public domain; 210 Colonel Light Gardens Historical Society; 211 Mary Evans Picture Library; 212 Mary Evans Picture Library; 213 Studio Books, London; 214 ontheinside.info; 215 Hamburg's Grünes Netz, created by Behörde für Stadtentwicklung und Umwelt (BSU), Kartengrundlage: Landesbetrieb Geoinformation und Vermessung, Hamburg 2013; 216t Foster + Partners; 216b NIR ELIAS/Reuters/Corbis; 217t Topic Photo Agency/Corbis; 217b Topic Photo Agency/Corbis.